Technically Food

Technically Food

Inside Silicon Valley's Mission to Change What We Eat

Larissa Zimberoff

Abrams Press, New York

Published in 2021 by Abrams Press, an imprint of ABRAMS.

Library of Congress Control Number: 2020932351

ISBN: 978-1-4197-4709-0
eISBN: 978-1-68335-991-3

Printed and bound in the United States
10 9 8 7 6 5 4 3 2 1

ABRAMS
The Art of Books

195 Broadway
New York, NY 10007
abramsbooks.com

Contents

Introduction

Why Me?

I first started thinking about the macronutrient content in food when I was twelve, after I peed in my pants in school. I was thirsty all the time, too, but I didn't make the connection. Mom took me to the doctor for an earache and casually mentioned my recent spate of mishaps. A urine test revealed that I had type 1 diabetes. That same day, I was checked into a hospital, and a nameless diabetes educator taught me how to calculate the grams of carbohydrates in a meal. (It's as hard as it sounds.)

The result of these mental gymnastics determined how many units of mealtime insulin to inject with a needle. (It's as bad as it sounds.) When I got it wrong, the physical repercussions ranged from being coated in sweat to feeling like I was moving through quicksand. Simple pleasures, like cupcakes at birthday parties, potato pancakes at Hanukkah, and Grandma's famous challah French toast on weekends became fraught calculations.

In my world, then, food is only as good as its primary building blocks, which are carbohydrates, protein, fat, and fiber. When I eat an apple, what I'm really eating are the macronutrients it's composed of. I choose green apples because they are typically less sweet than red apples. Less sweet relays to my brain: fewer carbs.

As the understanding of my condition improved, the constant oversight lifted. Exercise was as necessary as learning to like coffee black and chocolate dark—key skills for my people. My knowledge gives me an edge: I understand food on a molecular level. I think of this as my superpower, and also the difference, for me, between life and death. I see through food.

Like a Russian doll, my approach to sustenance is a nesting stack of questions: What time of day is it? How much will eating affect my blood sugar? Am I going for a walk after I eat? How much of what I'm about to eat is processed or packaged? I read a lot of labels.

When I spot a new food in the store, I look at the nutrition facts panel before I pour it into a bowl. This label is one of the most reproduced graphics in the world, yet very few people pay as much attention to it as I do. Only 31.4 percent of shoppers looked at the label "frequently," says a 2018 study published in the *Journal of the Academy of Nutrition and Dietetics* of nearly two thousand young adults. While a label can hide as much as it reveals, it's still our most valuable resource when choosing what to put in our mouths.

In my thirties, I realized that most of the world couldn't be bothered to see food the way I did. When I began covering the food-tech industry, I felt that this was my special contribution as a writer. Built upon this framework—that food is only as good as its components—is my decade-plus background working in high tech, an experience that fast-tracked me through the startup world. The frenzy in food investments I see now feels eerily similar to the first wave of the Internet.

Covering the world of food tech startups means I'm surrounded by (mostly) young entrepreneurs who are certain they can "make the world a better place." Their confidence is validated by the millions of dollars they raise—a sign that they're either brilliant or on to something big. I want a triple bottom line: good for me, good for the environment, and good for business. This book began when I asked myself: "What do we gain and what do we lose by embracing a future of lab-made food?"

The current wave of food companies claim to be mission driven. They want to better our world with futuristic processes. They hope to reverse climate change. They want to end animal suffering and the attendant damage to the planet from industrial agriculture. But they still want to make money. Capitalism is pulling the levers. Now, companies like Tyson, Nestlé, and General Mills (just a few of the legacy brands that I sometimes shorthand to "Big Food") are feeling the pinch from declining profits—their dated portfolio of products is no longer winning new generations of consumers—but they're not going to allow themselves to be left behind. I want to believe everything that New Food (my name for the startups in my book) assures me, but are they following the same path as Big Food? These are the very same companies that have fed Americans for decades, raked in profits, and given us in return higher rates of obesity and disease.

The tension of my health being tied to capitalistic companies that want to make a profit is growing. It affects people like me with diabetes, my sister-in-law who has celiac disease, my best friend's three-year-old who loves sweet foods, food-insecure communities, elderly populations, and unhoused people. Food affects everyone. Now that the world population is in the billions, and our natural resources are showing signs of breakdown, we want to know: Can we be healthy, respect food traditions, and save the environment all at the same time? It's not too much to expect.

Why Now?

Famous for getting seat belts into cars, longtime consumer advocate Ralph Nader is also known for cleaning up baby food in the seventies. Nader's problem was that manufacturers were putting additives—modified food starch and MSG (monosodium glutamate)—into infant formula. Companies weren't doing this for babies' health, which could potentially be harmed with high levels of glutamate (the primary amino acid in MSG);

they added it so that it would taste better to mothers, prolong its shelf life, and improve solubility, which makes it easier to mix. At issue for Nader was that the Food and Drug Administration (FDA) wasn't proactive, it was reactive. The burden of discovering problems in our food supply, he said, was left to researchers outside the industry. "One of the enduring characteristics of the food industry is its penchant to sell now and have someone else test later," said Nader.

It's been fifty years since that regulatory fight. Eventually, the FDA banned MSG from baby food, but to hedge its bets and keep companies happy, it declared it "fit for human consumption but not necessarily by infants." The other moral of the story is that this is happening with the New Food companies. The FDA still works reactively, and food companies are still getting away without proactive confirmations of food safety.

Recently, a new category of toddler milk launched in reaction to a slump in sales of baby formula. The new milk has a long list of ingredients including frowned upon substances such as corn syrup, palm oil, and polydextrose—a form of fiber used to improve mouthfeel. And baby food is just one small category of processed foods. Harmful gunk is everywhere: synthetic food colorings, saccharin, pyridine, and many more. All are still in use today, despite the fact that the FDA has all of them on a list for removal because they're carcinogenic. You might wonder why they're still in use. It's because the FDA gives companies years to reformulate and get rid of banned ingredients. In the meantime, there is no product recall and you can readily find the toddler milk on Amazon.

We expect that the foods we eat are the safest they have ever been. In many ways, they are. I won't deny that our regulatory system basically works, but our global health is failing, and that's largely because of the prevalence of the American diet. It's time for us to scrutinize our old habits as we consider shifting to these New Foods: milk that doesn't come from the udder of a cow, eggs that aren't laid by chickens, and shrimps that don't swim in the sea. Tomorrow's foods depend on highly trained scientists, many of whom have crossed over from the field of medicine. Tissue and cellular biologists, analytical chemists, food scientists, and engineers are collaborating on how to create food that they claim will benefit the world. But to feed billions, we need a supply chain that scales. To make food from literally almost nothing—yeast, bacteria, single-cell organisms, carbon emissions—we need industrial systems, which depend upon crops like sugar and corn (the same ones that are in use today), and we need chemical nutrients like insulin, growth hormones, and amino acids. If our health is failing us on the current industrial methods, shouldn't we be looking for ways that don't perpetuate that same framework?

As food moved from farm to factory over the course of the twentieth century, the conventional wisdom held that we didn't want to know "how the sausage is made." That the slaughter of animals for human consumption was a necessary evil, but one that people didn't want to think about. In the early 2000s, it seemed that transparency was beginning to draw open the blinds. Chef Dan Barber and author Michael Pollan told us that the quality of our food was crucial, and that flavor was a sign of a rich ecological legacy. Millennials were reported to want food from mission-based companies that they could feel good about. Greater accountability in our food systems, the uptick in special diets, the boom in physical fitness, and new nutrition research brought about welcomed change.

Writers have taken on this task before, perhaps none more impactful than Michael Pollan in *The Omnivore's Dilemma* (2006). After reading it, tech workers ditched their corporate jobs to start farms; consumers became more invested in what they were buying. Big Food felt the heat. But Pollan didn't get there on his own. Frances Moore Lappé's book *Diet for a Small Planet* (1971) began this dialogue in earnest. "The most wasteful and inefficient food systems are those controlled by a few in the interests of a few," she wrote. This sentiment still rings true.

When I began covering the intersection of food and technology in 2015, I raced to keep up with rapid launches in the food space while I waited to see if any founder would mention these writers and their lessons. They did not. Similar to Nader, Pollan was deeply concerned about our reliance on a food system we didn't understand. He wrote: "It is very much in the interest of the food industry to exacerbate our anxieties about what to eat, the better to then assuage them with new products." In his book, Pollan was referring to our "bewilderment" at the overflowing shelves in the supermarket, and the endless parade of new foods. That was 2006, but little seems to have changed.

Because of my vigilance over my own diet, I look for ways to lessen the mental load that comes from living with diabetes. I've tested various eating programs, including Whole30, intermittent fasting, and plant-based keto. Through them, I've discovered that the fewer processed foods I eat, the less effort it takes to regulate my blood sugar. It's a simple rule that took me decades to learn. We're in the midst of a huge food experiment. New Foods are replacing whole plants with processed plants and traditional proteins with analogue proteins. Which versions will be best for our bodies, our guts, our health? Are they as good on the inside? The journey for me within the pages of *Technically Food* was to uncover the origins of these mysterious foods and demystify them through my reporting. In my role as food experimenter, I will find out for you.

I wrote this book to help people who love food, like me, get a little more science-savvy. The future foods I've investigated may or may not help reverse impending environmental doom, and it remains to be seen whether they offer a more holistic and pleasurable way of eating. In this moment of inflection, as our diets shift from animals to plants and simple to scientific, let's not lose the clarity we've worked so hard to achieve. My hope is that this book enables more conversations and shines a spotlight on how exactly the sausage is made. Even when that sausage is made from flora and fungi.

Chapter 1

The Future Food That's Always in the Future

Prehistoric Greens

Science fiction writers love to speculate about the weird and revolting things we'll eat in the future. Algae often tops that list, lending dystopian worlds their distinctly fishy undercurrent. Consider some of the science fiction meals of the 1950s and '60s. In *The People Trap*, the world subsisted on "processed algae between slices of fish-meal bread." In *The Space Merchants*, meat substitutes were fabricated from kelp fed by New York City waste, plankton from Tierra del Fuego, and chlorella from Costa Rica. In *Make Room! Make Room!*, we read of a diet subsisting of crackers thinly spread with margarine, whale blubber, and chlorella. That novel was the basis for the 1973 film *Soylent Green*, which predicted dying oceans, depleted resources, and year-round humidity—environmental phenomena that we now see in the daily news. Forty years later a product called Soylent, which originally included algal oil, became one of the first truly weird, out-there foods. Made by someone who viewed eating as time-consuming and annoying, the cloyingly sweet meal-replacement shake is a good example of why coders shouldn't dabble in food.

But the premise of algae (and its larger cousin kelp) is that it's abundant and nutritious. I take spirulina supplements for their omegas, I eat kelp jerky, and I order wakame and hijiki seaweed salad when I spot it on a restaurant menu. I love the idea of these things—they are prehistoric greens. Algae are found abundantly in the ocean and are the primary carbon source for marine protein. If you eat fish,* you are indirectly eating algae. And while fish are a good source of omega-3 fatty acids—eicosatetraenoic acid (EPA) and docosahexaenoic acid (DHA)—they get those compounds from algae. Don't like fish? You can get these crucial fatty acids straight from algae, too.

Algae (singular: "alga") are marine organisms that have been around for more than a billion years. Archaeological evidence shows that algae have been part of the human diet for thousands of years. Spirulina, a commonly found blue-green alga, has been consumed as a food for centuries in central Africa. In Chad, where it's known as *dihé*, the tiny organism is harvested from Lake Kossorom, dried in the sun, and used in meat and vegetable broth. In Mexico, more than four hundred years ago, it was harvested and used as a dry ingredient in a cake known as *tecuitlatl*.

Algae have been sent into space to feed astronauts, although not very successfully—the astronauts hated it like kids hate spinach. NASA had the right idea, but it needed to work on its recipes. Algae offer up the possibility of feeding humans and animals on terra firma without the environmental damage caused by intensive farming and fishing, but, like science fiction writers, many people don't, in fact, think much of the idea of eating it.

* There are companies working on microalgae-based fish feed for aquafarming—land-based fish production, but the cost is still too high to gain widespread adoption from the industry. This may be the most viable path toward algae as food.

But startups continue to chase algae for its mostly unfulfilled promises. It's a protein super factory: Some species are so high in protein that meat, milk, eggs, and soybeans pale in comparison. It's growth agnostic: Algae grows in water, either man-made or natural, and can, in a sense, be told what to do with few nutrient inputs. It's pretty: It can give candy a blue or green color. It can fuel your car. Algae has long been made out to be a miracle substance, luring entrepreneurs and scientists alike. But few have been successful in actually bringing it to market. Algae has long been a substance that fuels the world—and yet its realized potential for humans is always just around the corner.

Algae: The Game Changer

The plant kingdom is big; the kingdom of algae is vast. Although there is little singular agreement on exactly what an alga is, scientists have identified about one hundred thousand species. Don't even get them started on classification. Of those, only around two hundred have been introduced into the human diet. Algae are simple organisms. They use the sun's energy to multiply rapidly. When compared with other, similar aquatic organisms, algae are just about the most productive at making fat, protein, and carbs. They have been heralded as a source for an almost endless list of modern needs, including biofuels, food, fertilizers, and natural dyes. Their flexibility in output, and their comparative sustainability, are why algae are considered the most highly efficient microorganism.

I wanted to believe in algae.

My research took me to interesting nooks of the food world to learn about the potential. I found red algae, which is a macroalgae—this just means big—traditionally found in the sea. Red algae has seven thousand species and counting. On the blue-green spectrum is cyanobacteria, which includes chlorella and spirulina, perhaps the most commonly found supplement that has yet to take center stage. There's euglena, a single-celled microorganism that has both plant and animal characteristics. It's not really algae, but close. And finally, I found duckweed, which isn't algae either but has great nutritional potential. They're all the same family despite the ambiguous grouping. These examples are all light-harvesting aquatic organisms, generously defined.

Science writer Ruth Kassinger, in her book *Slime* (2019), writes that "there are more algae in the oceans than there are stars in all the galaxies in the universe." Algae are doing the heavy lifting of supporting life on Earth. Fifty percent of the oxygen we breathe comes from them. Anything that harms algae is very bad news. Two thousand nineteen was the second-hottest year on record. And between 2015 and 2019, the growth rate of CO_2, which remains in the atmosphere for centuries, and in oceans for longer, was nearly 20 percent higher. This increase in CO_2 leads to a rise in pH levels called ocean acidification, which can harm many creatures in the water. Algae, on the other hand, may benefit from increased levels,* and there are studies looking into whether growing seaweed can slow this rise in pH levels.

"I think there's a big promise in microalgae," Asaf Tzachor said to me on the phone. A researcher at the Centre for the Study of Existential Risk (CSER) at the University of Cambridge, England, Tzachor looks at critical, life-supporting systems, not just food, and how they might play out under certain stressors. Can microalgae be affordable as well as safe and nutritious? Are microalgae farms plug-and-play systems that can be taken to other countries? Is there any off-the-shelf technology that can be utilized? He told me that "ticking all of these boxes is difficult."

For Tzachor, algae are a peerless crop. You can eliminate the natural components with enhanced artificial components. You can replace sunlight with LEDs for growing; you don't need arable land; and brackish water is fine. You can stack up growing environments and put them in cities. You can adjust the spectrum on LED tubes to optimize the growth rates. Algae could supposedly alleviate hunger and cross-cultural boundaries . . . but that hasn't happened yet. It's this long list of potential that

* Harmful algal blooms, a large swath of uncontrolled growth in both fresh water and our oceans, are of concern due to environmental changes as well, but the nutrients that typically fuel these blooms are phosphorous and nitrogen.

continues to fuel a kind of algae fervor, and that keeps the green slime bubbling up in any conversation about food system overhaul.

Yes, but . . . Can Algae Be Delicious?

Although seaweed snacks are beloved in Asia, US adoption, even with progressive millennials, has yet to move much further than crisp sheets of nori. One of the few startups working diligently to make algae delicious is Nonfood, based in Detroit. Having begun as a bit of a conceptual art project spun into a small business, Nonfood makes an algae-based snack bar called Nonbar. Like space food, or backpacking rations, version 4.0 arrived in the mail wrapped in silver foil. It was small, dark green, and flecked with bits of roasted fava beans. The ingredients were promising: lemna, chlorella, and spirulina. In addition to 7 grams of protein, one bar includes 27 percent of your recommended daily iron intake, 100 percent of vitamin A, and 438 mg of omega-3s, including alphalinolenic acid (ALA), DHA, and EPA.

I ate my Nonbar while walking to the library, biting down and chewing slowly. It was unusual, a little floral, a little savory. Not very sweet. I couldn't quite say I liked it. But I kept eating. A little like dating, after my third bite I was in, and I began to relish it, to the point that when I finished, I was sad to see it go.

Nonfood founder Sean Raspet is an artist who spent a few years working as a flavorist at Soylent. Raspet told me, "My art is about redefining boundaries between what we think of art and the mass economy. Nonfood fits into that." Although I embraced Nonbar, I wasn't sure how eager the world would be for designer food—bars that were undeniably good for us, but that came wrapped in an Andy Warhol art vibe.

Elliot Roth is another founder who has spent years, probably more than he wants to admit, attempting to harness algae for our diet. Spira, Roth's startup, began in Virginia in 2016. The first product was a kitchen gadget that produced fresh algae at home—like a coffee maker of sorts. It didn't go far. Then Roth made "live" spirulina tea. When I asked him to ship a bottle to me in New York, he said it wouldn't make the trip: "The shelf life is short." While tea sales were sluggish, Roth spent time in the lab tinkering with other ideas for his spirulina species.

The "easiest" way to grow algae is in giant, open ponds the shape of small racetracks. To keep the algae from going stagnant, a wheel moves the water around while nutrients are piped in. Every two days or so, when the material has reached saturation, the algae are harvested. "There are thousands of spirulina farms around the world struggling for business," Roth said. These farms are primarily in hard-to-reach rural locations, and clients aren't easy to find. Spira can sign contracts for better prices and offer them steady business. His startup works with farms in Indonesia, India, Thailand, and Mongolia. "We'll use the network effect," he explained—a way to harness algae farms to grow enough material to supply his growing customer base.

Unfortunately, open ponds are affected by air pollution and weather, and producers are experimenting with other methods. The options include growing algae inside giant fermentation tanks, or in photobioreactors, a maze of enclosed glass tubes that continuously flush nutrients, creating a constant biomass. Both methods depend on energy from the grid, which makes them less sustainable than using sunlight, and means they would require a significant amount of money to get production up and running.

Eventually, Roth gave up on the tea, but not before figuring out how to engineer his green spirulina species into spitting out a vibrant blue compound for food manufacturers. After three years of poking and prodding algae to become a food in its own right, Roth was now creating a food additive for what he felt could be more impactful: dye.

Not Food, but How About Hue?

In 2019, Roth met me at Stonemill Matcha, a sleek café in the heart of San Francisco's Mission District. It was our first in-person meeting. The café door tinkled, and Roth blew in. He sat down, took his knit cap off, his hair shot out, he reached into his pocket and pulled out a clear screw-top container filled with bright blue dust. "We call this 'Electric Sky,'" he said, handing it to me. "It smells like cheese."

I hovered my nose over the container, and then pressed my finger into the fine powder. He was right; it did have a Cheeto-like quality, but the shocking blue color played mind games with my taste buds. Like when Heinz launched purple ketchup, it didn't compute. He tapped it into his matcha latte, and I did the same. "Why not?" I said. "Polyphenols, right?" I swirled the blue into the green tea, and admired the tie-dye effect. Then I wondered whether I had ruined my overpriced drink.

The blue was brilliant, like cobalt painted on fine Chinese porcelain. In large amounts cobalt is toxic; ditto the petroleum-based blue dyes used widely in today's food manufacturing. It was what made Roth focus on the somewhat unnatural color. "People don't want petroleum-made ingredients" in the food supply, he said. Since 2008, the Center for Science in the Public Interest has urged the FDA to ban several food dyes. Mars, maker of M&M's, spent years and lots of cash developing a natural blue extracted from spirulina. In 2016, the *New York Times Magazine* ran a story titled: "Can Big Food Change?," in which food and farming journalist Malia Wollan reported on Mars's work to use a blue color derived from spirulina, and the algae industry's excitement to be the new food-dye replacement.

In the story, Wollan wrote: "The enormous amount of money and effort researchers at food companies are putting into the development of highly saturated natural colors is partly an effort to maintain the status quo, the legacy of products through time. But they also know that, biologically speaking, deep, contrasting colors are more attractive to people than mere nostalgia for a specific candy."

In 2016, Mars announced that it would cut out artificial dyes in all of its food products for humans over the next five years. Despite that, naturally colored M&M's have yet to make it to market. If you read the back of an M&M's bag, you'll see Mars is still using E133, otherwise known as brilliant blue FCF. (Always be skeptical of a food that goes by numbers.) I reached out to Mars, but did not get a response. "Mars has been burned so badly by algae," Roth said, referring to the money Mars spent on its algae research. "We're being really cautious because of that." I marveled at the blue powder in my hand. Spira had done it, or so it seemed. It had made a vibrant blue pigment from spirulina.

Spirulina and Chlorella have long been grown for the supplement industry because of their health benefits. But they're still considered exotic ingredients. In spite of Tzachor's high esteem, algae are limited by technology, which is costly. Plus, they're tiny and touchy to grow. Get something wrong—temperature, amount of light, nutrients—and they'll die.

While I stirred my matcha latte, Roth took another sample jar out of his backpack. He handed me a jar with irregular chunks of what looked like dirty white chalk. "Like a spirulina protein isolate?" I asked. He nodded, and told me his little engineered algae was 48 percent protein. To get to blue, and chunks of protein, Spira went into the lab and engineered algae that way. Another way of saying this is that it's been genetically engineered (GE), or genetically modified (GM). In 2018, the USDA agency landed on the word "bioengineered" for food labeling laws in order to describe ingredients like what Roth is making in the lab. You'll also see this listed as GE.*

In 2019, Roth moved Spira to San Pedro, a port town near Long Beach, California. The five-person team at Spira works out of a lab inside a modular shipping container in a parking lot. The startup is now part of a bigger marine incubator called AltaSea.org. A City of Los Angeles initiative to support the so-called blue economy, Alta Sea works to harness the world's oceans for economic growth.

Spira is small but hopeful. In 2021, Roth plans to fundraise to support its growing customer base. One customer is Gem, which is using spirulina in a daily chewable vitamin; 101 Cider House is making two ciders with Spira's colorant; Raw Juicery makes "mermaid lemonade"; and Noma, widely considered to be one of the top restaurants in the world, experimented with it. While most offices closed during the pandemic,

* The first GE salmon is expected to be available to seafood distributors in 2021 from AquaBounty. It took almost three decades to get approval from the FDA.

Spira's bitty lab pumped out samples. "They're bored and will experiment," Roth said. "And now we have a pipeline of sales."

Roth has grand visions of being the blue in Pfizer's Viagra pill, and in the latex paint used by the Blue Man Group—all the really important blue stuff—but "big companies want to see the tech de-risked before they take a chance," he said. Spira hasn't fixed its blue to withstand cooking, heating, and cooling, but it's close, and it recently added red to its lab work, which can be extracted from Rhodophyta—or red algae.

This color angle is picking up speed. San Diego–based Triton Algae is growing Chlamydomonas reinhardtii, typically a green alga, in its pilot facility. In the lab, Triton dials up the iron, or heme, which lends the algae a red color and gives it a meat-like flavor, according to the startup. Color and flavor make Triton's algae-based ingredient attractive to plant-based meat makers. In its fermentation process, Triton stimulates the receptors for the green color of its algae, which would traditionally include chlorophyll. Because of Triton's process, which they don't elaborate on, the algae turn red, producing heme precursor compounds. It's this heme, or iron, that lends traditional red meat its taste. Because red isn't ideal for everything we eat, nor is a meat-y taste, Triton is working on a colorless, flavorless version.

Is Algae's Future in Seaweed?

Microalgae won't feed the world anytime soon, but seaweed (a macroalgae) looks more promising. It's also the quickest path to the supermarket. Seaweed is a staple in Asian diets, but in our everyday foods it's mostly used as a binder, thickener, or gel. Carrageenan is the most widely used additive from seaweed, and is formulated into common foods including ice cream, plant-based milk, yogurt, candy, and, yes, baby food.*

* Carrageenan is outlawed in the European Union, but in the US it can still be formulated into baby formula. The Center for Science in the Public Interest (CSPI) recommends "caution" in the intake of carrageenan.

What is commonly overlooked is that it is a valuable source of protein. In the range of red, brown, and green seaweed, red leads with a protein content of almost 47 percent of its dry weight. In a head-to-head comparison with soybeans, one of the cheapest crops to grow, red seaweed produces five times more protein per acre. This makes it more valuable and more sustainable. Because the healthy ecosystems supporting our arable land—land that can grow food—are depleted by industrial farming, savvy entrepreneurs are turning to the ocean for its existing seaweed supply chain.

Compared with commercial overfishing in our oceans, and the subsequent decline in many ocean species, seaweed can be grown and regrown with relative ease—as long as the waters aren't polluted or too warm. Whereas soil-based industrial farming depends on fertilizers, the ocean has a steady flow of its own nutrients including nitrogen, one of the key building blocks of protein. Plants also need movement, which in the ocean comes from wind and waves that stir the water up. Seaweed provides all of the essential amino acids we need to survive—including glutamic acid, glycine, and B_{12}—but it's the protein that may make it immensely more valuable.

"Seaweed entrepreneur" isn't the catchiest phrase, but it still represents big business. In 2019, the global seaweed market was valued at more than $59 billion. In 2020, it was projected to grow another $6 billion. That's enough to entice some, including Beth Zotter and Amanda Stiles of Trophic, a food-tech startup based in Berkeley, California. I stopped by their lab on an overcast day. From the parking lot, I could see out to a small lagoon, and beyond that, the San Francisco Bay. Inside the office we sat in a windowless conference room. Stiles set out a tray with seaweed under evaluation. Curly, and spiky, red and pink. I stuck my nose into each bowl and inhaled their briny scent.

"We think that seaweed has been ignored and overlooked," said Zotter. According to Trophic, the red color molecule, which is tightly bound to the protein, gives seaweed its rich, savory umami taste. Zotter told me that some varieties—especially dulse—taste like bacon. The high protein and the color could be a boon to the production of plant-based meat. "Seaweed is also cheap. You can get it for ten to fifty cents a kilogram," Zotter noted.

Prior to Trophic, Zotter worked for a Japanese company that was investigating the use of macroalgae to produce biofuels—a money pit that has swallowed many startups. The failure (so far) is finding an economical way to grow seaweed or algae at a scale that also supports investment dollars. Stiles, Zotter's co-founder, worked at Ripple Foods and created scalable processes for isolating protein from plants—but for food. "At Ripple, I was trying to completely purify protein, whereas here we're excited about everything that comes along with the protein," she said.

Food is elemental for Trophic. "Our goal is to beat the cost of soy protein on both price and scale," said Zotter. At two dollars a kilogram, soy is almost too cheap to compete with. This means Trophic has its work cut out for it. Somehow, during the early months of the pandemic, the tiny team of two made progress. Stiles built out a lab in her garage complete with a centrifuge, and Zotter applied for research grants, and wondered which product to develop first.

Last January, the team found a way to extract a 50 percent protein concentration; a second centrifuge is on order; and one ton of red seaweed is waiting to be processed at a nearby USDA facility in Albany, California, which will help them scale protein extraction and test new drying techniques. Once the facility reopens, they can send protein samples out to dozens of commercial companies eager to test it in their plant-based products. "We're raring to go. We're just waiting for 'shelter in place' to be lifted," said Zotter in July 2020.

The Seaweed Diet

We take prescription drugs once in a while, but we eat several times a day. We know all the details about the safety of our drugs, but food companies can add ingredients whose safety has only been established by their own experts. Food-tech startups tend to follow one hierarchy for their products' reasons for being: It's good for the planet. It saves animals. And it's better for humans.

There is broad consensus on what constitutes a healthy diet—whole foods, the Mediterranean diet, plant-based—but nutrition science continues to waffle on other key areas like fat, red meat, and carbohydrates. Eating algae as part of our diet rarely makes an appearance on the world stage of eating guidelines, so . . . is it healthy? Well, it's a tough question to answer. In a 2017 study in the *Journal of Phycology* (the study of algae), researchers wrote that assessing the algae benefit to human health was "more tenuous." This is because too few studies have been done, and we're all uniquely different in how we metabolize foods. But if I consider the health and longevity of cultures that have eaten seaweed for centuries, I for one will continue to turn my teeth green by eating more Nonfood algae bars.

In Asia, studies have pointed to seaweed consumption as helping to prevent cancer. A study in the United States done by Francis J. Fields et al. in the *Journal of Functional Foods* in 2019 looked at the effects of microalgae on gastrointestinal health. The research came out of Steve Mayfield's lab at the University of California, San Diego. Mayfield is one of the more knowledgeable people in the algae industry, having spun off two algae-based companies, including Triton, while running his university lab. Participants reported less gastrointestinal distress, but it was a small sample that was self-reported, and larger studies are needed in order to confirm these results.

Consistency is key in commercially available food, but in the natural world it's hard to come by. Algae that's grown in the ocean has natural variation across species, seasons, and coastal environments. Another area of variance with algae is our limited understanding of its bioavailability—how well algae macronutrients are absorbed by the body and how its nutritional content interacts with our own metabolic process. While we think algae is good for our health, we don't know exactly how it's good for our health.

On the nitty-gritty of sustainability, the evidence is clearer. In a report coauthored by Tzachor published in the October 2017 issue of *Industrial Biotechnology* titled "Cutting Out the Middle Fish: Marine Microalgae as the Next Sustainable Omega-3 Fatty Acids and Protein Source," the researchers examined the amount of land and water needed to produce an equal quantity of essential amino acids from various standard foods that we eat. Beating chicken, beef, and peas was marine microalgae, which "reduces land usage by over 75-fold, since no fertile land is required, and lowers freshwater usage by a factor of 7,400."

One protein not featured in the *Industrial Biotechnology* report was one that uses (possibly) the least amount of land. That's protein from air, a concept that first appeared in the science fiction novels of Jules Verne more than a century ago. NovoNutrients, of Sunnyvale, California, is taking carbon dioxide from methane emissions—the gases that we want to minimize in our ozone layer—to create protein that can be formulated into fish feed for aquaculture farming. NovoNutrients is embarking on a pilot with Chevron. Another startup is Air Protein, which proposes to turn air into protein, and the resulting protein powder into edible foods. Outlandish-sounding, and I'm told that it has already made an air-protein "chicken" analogue. No word on how it tastes. The Berkeley-based startup raised $32 million in January 2021. A third is Iceland's Solar Foods, which uses renewable hydroelectric energy to achieve the same end.

Critics of this "out-there" idea are not hard to come by. Mark Jacobson, a climate scientist at Stanford University, thinks that taking carbon dioxide from the air to produce food will be too energy intensive. "Food isn't just carbon, it's also hydrogen and nitrogen," he said. "You need energy to create the food. It sounds great, but it's a gimmick. We don't need to take carbon out of the air, we need to stop it from getting there in the first place." Jacobson's life work is to see coal plants around the world replaced with sources of renewable energy. Making food with a coal plant only creates more carbon emissions from that same bad coal plant. When we think about shifting away from growing food and raising animals on farmland, and producing more food in manufacturing plants, we can't forget about the need to build the infrastructure, supply the water needed to grow the food, and connect the energy to run the plant.

Algae has the potential to feed the planet—if only enough consumers can be enticed to ingest it, or a startup can successfully find exactly the right form or function. Is it protein, flavoring, coloring, or something as yet discovered? Roth of Spira agreed on the seemingly infinite applications and possibilities. "It's a tenet of the algae community. Algae can do so much," Roth said. This fervent belief extends to outliers in the algae category, which is where we might finally get a successful case for extracting protein from tiny plants.

The Promise of Tiny Things

"Look for the leaning white mailbox," Tony Martens wrote over email. I turned onto a dirt road with potholes, coming to a stop near a mobile office trailer. In the distance were the hills of northern San Diego. Behind the offices I could see several tattered-looking, plastic-covered hoop houses. Martens walked out with an open and inviting grin, and welcomed me to their new digs: "We just rented these." The founder towered over me as he pulled up his jeans with his wrists. His excitement was palpable. Maurits van de Ven, his co-founder, walked out of a door with a wet head, holding a plate of broccoli and fake meat. I tried to determine who was the science guy and who was the business guy, but I couldn't tell. While Martens talked about how they got from Amsterdam to a dirt road in northern San Diego, California, van de Ven ate his lunch.

Plantible Foods grows lemna—a tiny, floating aquatic plant—not quite an alga, but a bountiful source of RuBisCO. In photosynthetic plants, RuBisCO is the enzyme responsible for the first step in carbon fixation, whereby carbon from the atmosphere is taken in by plants and converted into other forms of energy like glucose and protein. Lemna, which is used in Raspet's Nonbar, is 40 to 45 percent protein. RuBisCO is the most commonly found protein source in the world. Despite being eaten by birds and aquatic animals, in some places wild-growing lemna is considered a noxious weed—like kudzu—because it can completely cover a body of water and hinder the growth of other plants. But growing it for its protein, and for humans, has potential, according to these founders.

"The cool thing about RuBisCO is that it behaves like an egg white, whey, or casein," Martens said. "You can create cheese, dairy, or meat-like textures more efficiently, and at a much lower concentration than soy, pea, wheat, rice—you name it." The problem was that in the more common sources, "green leaves that we can chew," growers don't want to isolate molecules from foods that are already an easy sell—such as kale, spinach, lettuce. I was beginning to see the challenge. Waste streams from farms, like broccoli leaves or carrot tops, were another good source for RuBisCO—but getting a consistent, clean supply is an obstacle that would shift seasonally.

It says a lot about the level of excitement in an industry that two thirty-year-old entrepreneurs left Amsterdam (where they were surrounded by water), and moved to northern San Diego (mostly sand) to launch a business growing tiny aquatic plants (that need water), and which will somehow replace eggs in baking or milk in yogurt. The buzz in the food world—a mix of Gold Rush enthusiasm and activist sweat equity—is due to a mix of two things: investor wealth coming primarily from Silicon Valley hunting for the next unicorn, and earnest zeal for saving the planet. While our established food system had the brainpower, labs, and financials, big corporations had little motivation to look for alternatives. To believers in conventional or industrial agriculture, Earth's resources are endless. According to then-President Trump, climate change doesn't exist. Thankfully, there are many who do believe in climate change and are paying attention to the near constant wildfires, melting icebergs, and warming oceans that have inspired a whole swath of food companies with different goals. It's worth noting that Big Food is watching, and buying up New Food, which may either squash all this ingenuity and do-goodism or allow it to prosper.

Apple Green

Lemna is green like a perfectly polished Granny Smith apple. In the hoop house at Plantible, it floated quietly atop an oval-shaped pond as paddle wheels circulated the water and air breezed by. Plastic wall coverings kept the temperature warm and humid inside. A drip-drop sound somewhere added to the meditative feeling I had gazing down at tens of thousands of wall-to-wall, double-leafed plants. "It's hypnotic," I told the pair. They laughed. I wasn't the first to make this observation. "Can I taste it?" "Sure," they said. I dipped my forefinger in and brought it back up coated in watery green fragments. It looked like broken-up edamame. I put it in my mouth. It tasted like iceberg lettuce or the stem of a tulip, which I'll admit I've also tasted—watery and crisp.

"We've looked at basically every green leaf there is," said van de Ven. "From alfalfa to chlorophyll containing algae that all have RuBisCO. Then we looked into the duckweed space," he said. I took a moment to consider the "duckweed space." Once propagated, the organism keeps growing itself, giving Plantible a self-perpetuating supply chain with no downtime. We walked through a few hoop houses, which the pair inherited from a company that went belly-up after trying to make a go at growing algae commercially. (See: believers!) Then Martens led me over to the protein processing area—another room on wheels.

Plantible's frugality was the antithesis of the startup I worked for in the late nineties during the Internet bubble where we played Foosball and sat in overpriced Herman Miller Aeron chairs. In addition to scoring property with eighteen hoop houses on it, and leasing cheap pre-fab offices, Plantible has economized. Instead of fancy microfluidizers that cost the equivalent of a new car, they use a blender. "It's hard to compete with the Vitamix," said Martens fondly, slapping the gadget on its side. Once whipped up, the green slurry went into what looked like a swimsuit dryer, the kind that eats your drawstring in ten seconds flat. Inside the spinner, protein and fiber were separated. Then, with the use of heat, the chlorophyll (the green pigment) was removed; finally, the polyphenols (the flavor) were removed using activated charcoal. The end product is a white, flavorless protein powder that Plantible can sell to food manufacturers. As for the polyphenols and chlorophyll, the startup is attempting to find ways to sell them off—possibly to nutraceutical companies. For now, the extraneous flavor and color found in lemna remain unrealized waste streams.*

This economy, however, means that Plantible is stuck shipping small amounts it can produce in its own lab to eager companies that want to test its protein. Many New Food companies struggle with whether to supply ingredients commercially or make their own consumer-friendly products. For now, Plantible is focused on scaling up its protein production. "Every day I get, like, twenty emails asking for samples," said Martens. "And we're like, OK, we need to keep the samples for ourselves so that we can develop our own products." Yogurt was high on the list, but that's a competitive area of the supermarket. As an egg replacer it may have more luck.

Six months later I checked back in with Martens. It was July and the coronavirus was raging across the United States. Despite the pandemic, Plantible had closed a $4.6 million round of funding in April. The founders had rented RVs and were living in them on the property. The team had grown and was testing different species of lemna for growth rate and protein content. While Martens had finally swapped out his beloved Vitamix and swimsuit dryer for a colloid mill and centrifuge, Plantible was still producing less than one kilo a week. In 2021, they hope to have a pilot plant that can produce ten kilos a week.

When Pat Brown, the founder of Impossible Foods, first began making his now famous burger patties, he used RuBisCO as one of his ingredients. "It worked functionally better than any other protein, making a juicy burger," Brown said. The problem, Brown told *The New Yorker*, was that no one was making it at scale. The eager Dutchmen from Plantible are betting they can prove Brown wrong, and the R&D team at Impossible even has a small amount of Plantible protein powder to test.

* A similar example of waste streams in food manufacturing is calcium from eggshells that is lost in the production of liquid eggs.

To get to the point of creating protein for just one customer, for instance Impossible Foods, Plantible would need more of everything. "Let's say [Impossible] needed one thousand tons of RuBisCO protein per year. That will mean that we need to operate two hundred forty acres,* which represents 0.0003 percent of the soybean acreage in the US." These were all estimates, but for now Plantible is operating on a two-acre farm with only one acre covered in soothing, hypnotic lemna.

If lemna can succeed where algae so far haven't, perhaps the business model will help propel the entire industry? Whether it's as a protein source, or a new food colorant, algae is both promising and vexing, which brings me back to its complicated status for entrepreneurs. It gets investment dollars, but not nearly the same levels of the other food-tech in my book. Investors say they want to fight climate change, but the money rarely funnels down to floating green bits. In November 2020, Jeff Bezos announced the winners of a $10 billion climate fund. None of the winners are looking at the food supply. Nonetheless, like the passionate founders I met, I am hopeful for the future of algae. In a vote of confidence in algae as a climate-friendly solution, the US Senate added the organism to its 2018 farm bill. Upgraded from a supplement to a crop, algae was granted a range of support aimed at promoting its use as an agricultural product—from crop insurance for algae farmers to the establishment of a new USDA Algae Agriculture Research Program.

While our own protein sources languish in the familiar, cows' diets are proving far more adventurous. UC Davis is running pilots expressly looking at ways to introduce seaweed to cows' diets to reduce the amount of methane they produce. It has run trials with both dairy and beef cows, and the results are compelling. Preliminary results have been shown that adding a small amount of Asparagopsis armata (red seaweed) lowered enteric fermentation, aka cow burps, which is what releases methane into the atmosphere. Even a tiny amount works. A diet of just 0.5 percent

* Two hundred forty acres is equal to just under 182 football fields, or 3,722 tennis courts.

seaweed led to a 26 percent decrease in methane, and a one-percent seaweed diet produced 67 percent less methane.

Albert Straus, founder of Straus Family Creamery, and owner of an organic dairy farm in Marin County, California, received a USDA National Organic Program waiver to run a six-week experiment feeding his dairy cows a seaweed supplement from Blue Ocean Barns. Straus thinks that cows are essential to reversing climate change, and hopes to prove that by getting his farm and dairy to carbon neutral by the end of 2021. Whether an umami-rich diet changes the taste of steak or milk remains to be seen, but at some point in the future, one that isn't just in science fiction fantasy, the seaweed won't feed people but will feed livestock instead.

Chapter 2

Fungi

A Steak Substitute . . . and Flavor Enhancer?

From Batteries to Chicken Breasts

Up until a few years ago, few of us could pronounce the word "mycelium" (my-see-lee-um), but it was the basis of one of the first non-soy meat substitutes on the market. It's called Quorn. It was also one of the first single-cell biomass proteins made for human consumption—versus animal feed. This novel food manufacturing was first floated in the sixties, alongside rallying cries announcing impending protein shortages coming from an ever-growing world population—three billion was the alarming number at the time. After looking at thousands of soil samples, scientists eventually settled on the fungi *Fusarium venenatum.* Fueled by carbohydrates, the fungi could ferment in tanks converting into an edible "mycoprotein." In only a few days one gram turns into 1,500 tons. But it took ten years to evaluate the product for human consumption, and Quorn didn't launch commercially, first in England, until 1985. The niche product limped along for decades, some people reporting allergic reactions and gastrointestinal issues (the same way some of us have issues from eggs, dairy, and soy), but today things are different. In 2018, Quorn's "chicken" nuggets were the fastest-selling product within the meat-free category at Kroger, a supermarket chain with more than 2,700 US locations.

Mycelium is a relative of mushrooms, grouped into one of the estimated 1.5 million species of the fungi kingdom, neither plant nor animal. Imagine a wide-reaching root network below a one-hundred-year-old tree. Now shrink it so that the roots are very fine, like thousands of hanging threads from a branch. This threadlike structure is what mycelium looks like. In the forest, mycelia (plural: mycelium) feed on wood, soil, bugs, and other nutrients. Because mycelium breaks down its forest floor surroundings—like dead bugs and browned leaves—the fungus is often considered to be nature's clean-up crew. Fungi was presented as a more realistic protein source than algae back in 1957 in an article in *Economic Botany*. Both organisms often made the list of scientists peering into our nutritional future, but apart from Quorn's success, no one looked at commercializing the food potential of mycelium until recently.

In Boulder, Colorado, the founders of Emergy Foods understood how nimble mycelium are. Before they made "steak" with it, they made batteries. Tyler Huggins and his co-founder, Justin Whiteley, met at the University of Colorado. Initially the engineering students were turning mycelia into tiny batteries. The pair grew mycelium strains using nutrient-rich wastewater from local breweries, and baked the mass into a charcoal-like substance that could be fabricated into electrodes. What ingenuity!

Buyers didn't line up, so Huggins and Whiteley parlayed their research into food. They screened thousands of strains to find those suitable for production. "We selected for speed of growth, nutrition content, taste, texture, and carbon conversion," said Huggins. Eventually, they chose one from their microorganism library. They named their hard worker "Rosita."

"Pea protein is the hottest kid on the block, but it has *off* flavors," Huggins said. He's right. A lot of people are turned off by the still-too-pea taste, leaving plenty of room for contenders. In 2019, Emergy raised more than $5 million in funding—a large amount for a seed round. This included grants from the Department of Energy, which is interested in advanced manufacturing and global competitiveness; and the National Science Foundation, which is prioritizing sustainable food production.

Emergy grows its mycelium-based protein in fermentation tanks like what you've seen in a brewery. It begins as stringy threads in a beaker. The fungi grow in a mixture of nutrients, including sugar, nitrogen, and phosphorus—what Huggins deems as "all safe stuff." In the tanks, the mycelium thickens and fills up the tank. It does this fast. In eighteen hours, strands of mycelium fill up a thousand-liter fermentation tank. From that one tank, Emergy can create about eighty to one hundred pounds of finished product a day, a process that uses 90 percent less land and water than the equivalent factory farm.

Fermentation of foods is well established in our culinary history, but growing fungus is more complicated than just tossing it in a tank with some basic nutrients. Founders of food startups can be tight-lipped; most use the expression "secret sauce" to avoid answering questions about what exactly goes into making their cultivated food. Huggins assured me that Emergy's protein was minimally produced with no strong chemicals. "The back end is like cheese-making. We remove the water and form it into chicken breast or steaks. We try to keep it under five functional ingredients," he said. A few months later when we talked again, Huggins said that the final product was made of only three ingredients: mycelium, beets for coloring, and natural flavors—things like spices and nutritional yeast. I was impressed. Compare that with Beyond burger and Impossible burger, both of which have fifteen to eighteen ingredients, several of which are highly processed and far removed from their natural beginnings.

In 2020, I got to taste their first product, a mycelium "steak." To be more approachable, Emergy is going by the name Meati when it launches in grocery stores. Whoever achieves something that closely resembles a sink-your-teeth-in-it steak will be rewarded, because, in Huggins's words, "Whole cut is where it's at." According to the nutrition facts panel, one four-ounce serving was 22 grams of protein, 10 grams of carbohydrates, and hardly any fat. In comparison, a four-ounce steak from a cow has 13 grams of fat and 26 grams of protein. The mycelium steak was also high in zinc and provided more than 30 percent of my daily fiber needs. Numbers like these should please everyone. (Including me!)

Included in my Meati box was "chicken," "steak," and two bags of jerky. I dove into the jerky first. Reddish-brown in color, the jerky was an easy win. Savory and chewy, it was delicious. For dinner one night, I cooked up the "steak" in a small sauté pan on the stove. Frozen when I started, the chunk of meat didn't look like much, but soon it was searing up in the pan. I added olive oil and butter and watched it sizzle. When it was done, I placed it on a cutting board and with a sharp knife cut against the grain. The meat was pink inside and brown on the outside. I placed a strip in my mouth and chewed. The texture was phenomenal, and the flavor, while a bit heavy on the mushroom-ness, which I like, wasn't that far off. Huggins promised me that this version still had room for improvement. With a little more fat, and a more irregular shape, the steak could be extremely convincing.

The FDA states that if 20 percent of your daily nutritional needs can be met by getting nutrients—protein, fat, vitamin D—from a single serving of a certain food, then it can be considered an "excellent source." Protein made from mycelium falls in that camp. In addition to its protein, mycelium includes complex carbs and is low in fat. It's high in polyphenols, which have an antioxidant effect. It also has calcium and magnesium, and contains all nine essential amino acids for adults, and a protein digestibility score (PDCAAS) of .99, which beats beef at 0.92. A PDCAAS score is a way to evaluate the quality of a protein based on the amino acids present and our ability to digest them. An egg has a score of 1.0, but chicken scores a .95. The PDCAAS of algae, which can range from 40 to 60 percent protein concentration, earns a much lower score. Fungi have been in the human diet for centuries, and there may be less of a mental leap to add them to our dinner as an alternative source of protein.

To test his product, Huggins hit the road and shared it with chefs. He said it was an easy sell because of its low environmental footprint, its short ingredient list, and the ease with which chefs could tailor it to suit their cuisine. The first place you'll find the mycelium "steaks" may be at Michelin-starred restaurants in Los Angeles, New York, and Chicago. A three-ingredient meat analogue was a no-brainer for chefs. Eager to try it out on even the most challenging of vegan guests, chefs were instantly won over by the high-moisture form and malleable flavor. Not only did chefs want to serve it at their restaurants because of its minimally processed appearance, but they also wanted to invest. Both Grant Achatz of Alinea in Chicago and David Barber (chef Dan Barber's brother) invested in the company's $25 million series-A round that closed in 2020.

Unlike most other food tech founders, Huggins is neither vegan nor vegetarian. A competitive advantage for a startup hoping to create steak. "My standard for meat quality is high. I grew up in Montana and my parents have a bison ranch, but I believe in sustainability and lowering meat consumption." Since matching the flavor of red meat is tricky, I asked Huggins what he thought about his competitor Impossible's use of heme (a genetically engineered molecule) in its own burger, and whether Emergy would consider using GMOs. "I don't think heme is necessary yet," he told me, explaining that he wanted to see "what the public thinks" of the flavor before they added anything that was genetically modified, and only rumored to be necessary. "Until consumers demand it, I don't think we need it."

Using Mycelium to Improve Food

Just thirty-five miles away from Emergy Foods is another startup based around mycelium, only this one doesn't want us to eat it. Instead, MycoTechnology is using mycelium as a processing step for *other* proteins. Unlike Emergy, MycoTechnology has no plans to sell its own products in a supermarket. Its pitch to food manufacturers is that it has made a functional pea and rice protein blend that is more neutral in flavor, with less strongly vegetable aftertaste. MycoTech's claimed mission, to make our food system more sustainable and healthier, is an honorable one. Whether its ingredients can also make our food more delicious, and whether these ingredients will be formulated to make healthier foods, remains to be seen.

Before burrowing into the food tech world of fungi, my only relationship was with the mushroom caps that I sautéed in pans. The fruiting bodies of mushrooms—stem, cap, and gills—seemed utterly different from the structures belowground and inside the lab.

Although you may not have heard of MycoTechnology, it has twenty-seven awarded patents and forty-seven pending, which has helped raise more than $85 million in funding. In a vote of confidence from investors and Big Food, MycoTechnology moved into a brand-new 86,000-square-foot manufacturing facility in Aurora, Colorado, halfway between the airport and downtown Denver. Alan Hahn, the CEO, told me that each of its twenty-four fermentation tanks was in a constant state of brewing or being cleaned to brew again. The opposite of a craft beer operation, this seemed more industrial: two-dozen hulking steel tanks all churning out a single ingredient. Manufacturing plants like this are a hidden part of our food supply. When I visited the plant in 2019, I noticed a small team from Memphis Meats meeting with the CEO. The Berkeley startup was in town to look over the nuts and bolts of MycoTech's plant. Many of the New Food cohort will be relying on bioreactors—vessels that support the growth of living cells—and a process not unlike fermentation. MycoTech's was one of the first large-scale facilities built expressly for novel food production.

I first heard of MycoTechnology in 2015 when the company announced it had found a way to remove gluten from bread and pasta. The key was mycelium strains, which could break down the protein found in gluten. Pasta flour was fed into tanks with mycelium strains. This provided the fuel—gluten and some carbohydrates—needed to grow the strains. The resulting pasta made from this mycelium-treated flour wasn't 100 percent gluten-free, but it was close, they said. Another mycelium-based ingredient made by the Denver company was a powder called ClearIQ—a by-product of the fermentation process. The company says that ClearIQ, used in tiny amounts, alters eighteen of the twenty-five receptors on our tongue that sense the bitterness, or harsh flavor, in foods. It can make strong coffee less astringent or dark chocolate less bitter. Used in almost a hundred commercially available drinks—the specific brand names all protected by nondisclosure agreements (NDAs)—it can make a better black tea, something we may not have realized needed any help. It's also used in chocolate bars and CBD edibles—the kind Martha Stewart peddles. When I visited MycoTechnology, Hahn couldn't share any brands, but he told me I could find iced tea that used ClearIQ at almost any convenience store. Because the amount used is so minute, you'll never see it listed on the ingredient label, but you should know that it might be there under the auspices of "natural flavorings."

"Natural flavorings" can be found on almost every ingredient list. Nadia Berenstein calls this vague reference on our labels the "black box" of the food formulation industry. I first met Berenstein, who has a PhD in history from the University of Pennsylvania, when she lectured at New York University on synthetic flavoring for a group called the Experimental Cuisine Collective. The diminutive Brooklynite had me widening my eyes in near-constant surprise as she talked over the methods used to make our favorite candies. Around the room were paper bowls of Hot Tamales and Circus Peanuts.

The words "bitter blocker" may sound high-tech, but even table salt can counteract the effects of bitterness. Try it: Sprinkle salt on a grapefruit wedge. Crazy, but it works. How did a by-product of mycelium fermentation match the accepted use of salt? To better understand my knee-jerk reaction to unusual ingredients, I called up Berenstein to discuss these "black box" flavorings. As in much of nutrition science, there were two sides to the argument. "Flavor modifiers are tremendously exciting ingredients and are real advances in food and flavor science," she told me. They have the capacity to reach for the "Holy Grail of food that tastes sweet, fatty, salty, or bad for us, but using fewer bad-for-us ingredients," she said. On the other hand, there is a palpable shift in how willing the younger generations will be in giving free rein to food manufacturers. "The ways [ingredients] work raises suspicion in an era of being reflexively alarmed at anything that is chemical." And why should we trust the food industry? It's not like it's done a good job of building trust with consumers.

But on this trip to see MycoTech's new factory, my focus was on something much different.

The Burger Competition

Inside MycoTechnology's food lab, I was introduced to Savita Jensen, the most sociable food scientist I've ever met. "Here's a lab coat," she instructed, her friendly eyes blinking behind square-framed black glasses. She handed me plastic goggles. I towered over her, but we're both short. Jensen introduced me to her co-workers, who were bustling around doing other tasks. One perked up when she learned I was looking for a place to dine, and the group suddenly veered into discussions of where I should eat in Denver. Jensen suggested bakeries as she pointed me to the tools of their trade: scales, bowls, and baker's racks stacked high with ingredients. That day's agenda included an hour with Jensen to whip up my very own plant-based burger. The next day we would grill it up. Jensen's recipe, refined over months spent achieving the perfect burger, noted precise measurements, all the way down to 0.18 gram of salt. While my burger would be similar to versions sold at your local grocery store, Jensen said her recipe, using MycoTech's protein (called "FermentIQ"), would accomplish the same taste with fewer ingredients.

To make my own plant-based burger, I snapped on some blue latex gloves. Jensen placed a mixing bowl in front of me and showed me how to zero out the scale. Using thin plastic weigh boats, I measured out ingredients. The star of Jensen's recipe, as it is in almost every other non-red-meat burger, is TVP, or textured vegetable protein.* Think of TVP as the "ground" in ground beef. MycoTechnology buys pea and rice protein concentrates†

* TVP was invented in the sixties by Archer Daniels Midland (ADM). In 1991, the company trademarked the name. According to food insiders the company aggressively goes after other companies using its acronym. In Chapter 3, I'll discuss TVP in more detail.

† Concentrates are made up of a lower percentage of protein than an isolate, which also means it's less processed. I'll talk more about this in the next chapter.

from a manufacturer in the city of Chuzhou, in Anhui Province of eastern China, four hours east of Shanghai. The sole reason for combining pea and rice protein is to achieve a high PDCAAS score.

Even as a tech foodie, I knew nothing of a PDCAAS score until I began researching this book, and then it became a competition of who could get closest to one.* "PDCAAS" was what most founders spouted off, but a newer acronym had been released called DIAAS, or Digestible Indispensable Amino Acid. This newer framework focused on evaluating individual amino acid digestibility, whereas PDCAAS looks at results from the entire food. It's interesting information, but outside of health organizations focused on malnutrition and dietary guidelines, I bet the only other person who cares is a weightlifter training for the next Olympics.

While I measured out the remaining burger ingredients, Jensen added warm water to a bowl of TVP and used a rubber spatula to gently mix the kernels, which now had the appearance of edible packing puffs. In all, my burger contained MycoTechnology's pea and rice protein blend, vital wheat gluten (another protein), methylcellulose (a starch used for binding), beef, chicken, and umami flavorings (not from real cows or chickens but from yeast extract), beet powder (for the pinkish hue), burger seasoning, salt, water, and coconut fat. I stirred it with a spoon until Jensen made me use my hands.

When I mushed it between my closed fists, pink slime squirted through my fingers. "Keep going, tighter," Jensen cheered from beside me. After a few minutes, I could squeeze the mix in my palms, and when I released it I could see fibers stretching between my fingers. "See that," said Jensen with glee. I was more than a little impressed that it was so easy. After ten minutes of squishing, she prompted me to form it into a patty. Then Jensen plopped it on a tray and labeled it with my name on a piece of tape—like food in a communal refrigerator—and stored it in the freezer to chill. She announced, "Tomorrow we'll grill it up!"

I met up with Jensen in her food lab the next day at high noon. "Are

* Current holders of a PDCAAS score of one are casein, whey, soy, and eggs.

you ready for the burger-off?" She grinned wide. We would be grilling up three different burgers including the one I had made. The others were a high-fat version Jensen had already prepared, and a Beyond burger purchased at a local supermarket. "Yep. I'm ready," I told her. She handed me a lab coat. Beside me was a sensory scientist who would help us qualitatively assess the patty on metrics like texture, bite, chew, aroma, and flavor.

After frying up the three patties, we lined up at the white Formica counter and looked over our blank score sheets. In front of us were the three plain burgers—unadorned by condiments. The sensory scientist walked me through how to stay unbiased throughout the process. Talking and opinions were to be kept at a minimum. Before we began, I grabbed a plastic cup to spit my chewed meat into so that I didn't get too full. (It's the same trick vegan founders use when they want to try animal products as a necessary comparison point.) Chew, taste, then spit. Like tasting wine, it was sometimes hard to remember that the point was not to swallow it.

Over the next half hour we smelled, bit down, chewed methodically, moved the meat around in our mouths, pondered . . . and spit into cups. While different, the three burgers were also basically the same. The texture was a close match, and the fats delivered savory flavors to my taste buds. All three were a perfectly good approximation of red meat. It made me wonder if our obsession with red meat was an obsession with the familiar, and, most of all: nostalgia. Burgers are experiential. They're eaten at Fourth of July barbecues, on Sunday afternoons, with our parents as kids, at the ballpark. All we really needed was something that could stand up against condiments and a bun. Nail the texture and you'll score a ten. Of course, I'm oversimplifying things. Beyond's burger has millions of dollars invested in its formulation, and an assurance to investors that it was technology,* not food, that made it better than the real

* Most Silicon Valley investors want to see technology as a basis for why a certain company should be invested in. If it's just making food, mixing it up in bowls, and so on, then what's the reason for valuation or differentiation and how does it get to be a one-billion-dollar business?

thing—and an assurance to the public that it's good-for-you, wholesome food. Meanwhile, Jensen and I had knocked it off with fewer ingredients, less processing, and even less technology.

"There's a lot of copycat innovation—not a lot of real innovation," said Hahn over lunch. We were at an Outback Steakhouse—his choice but my guess was that he liked the quiet booths. More than ten years ago, when the CEO first met his MycoTech co-founder, a scientist working with mushrooms, Hahn was diagnosed with type 2 diabetes. He saw the research as having potential for people like him. By changing his diet to mostly plant-based, Hahn had been able to reverse his diagnosis. "My doctor told me I was the first patient to follow his advice and get off the meds," he said.

Hahn initially got into the mushroom business after seeing his co-founder use mushrooms to make coffee taste less bitter. They eventually passed on coffee because the logistics of coffee didn't point to profitability. Whether or not his processed ingredients are truly making better-for-us plant-based foods, health was always listed as Hahn's primary goal. Although it's certainly money, too. In June 2020, the company raised a $39 million series-D round of funding, and going public is something the CEO brought up in conversation a few times.

While looking over the Outback menu, I spotted a brussels sprout appetizer listed as containing 1,000 calories. It's not hard to assume that customers will order this item thinking they're being virtuous, even though frying the veggies in oil and adding bacon would cancel out all notions of healthful. Hahn and I, type 1 and type 2 in recovery, ordered salads. I may view the world through a more rigorous set of rules, but is it wrong to expect companies to protect those that don't go to my lengths? I appreciate that there are people like Hahn investing the time and money to introduce healthier ingredients into our iffy supply chain. And I want ingredients that are known, through centuries of human testing, to be food. I like Hahn, but I am skeptical of his highly processed mycelium powder that rides alongside the black box of "natural flavorings," and whose safety—crops are grown in the United States, processing is done in China, foods are manufactured who knows where—is, essentially, unknown.

One of my tour guides when I visited MycoTechnology was Rick Becker, the CTO. As we walked through the plant, he rattled off a list of food additives that he once managed for ingredient giants: crystalline fructose (made from corn, it's sweeter than high-fructose corn syrup), dextrose, starches, and beverage alcohol. I think Becker thought he was impressing me, but these aren't good things. In fact, they are the underpinnings of our highly processed, unhealthy American diet. Each time I interrupted Becker to ask a question, he answered slowly: "Now, Larissa, I'll get to that." He had a head of white hair and exuded authority, no doubt from his decades working behind the scenes of our food system. I liked him nonetheless.

"Our product is an explosive. If you get a lot of dust in the room, and you have the right atmosphere and you get a spark, you'll have a dust explosion," Becker said with a smirk. "Just like grain does." The first dust explosion from food manufacturing is on record at a flour mill in Italy in 1758. Three years ago, five people were killed in seven explosions. With that, I buttoned up my coat completely, tugged down my safety helmet, and straightened my plastic glasses. I'd seen enough episodes of *MythBusters* to know the drill. We continued on our tour.

Many of the mycelium strains that MycoTechnology has cultured come from shiitake mushrooms, but Hahn told me they had "over sixty different fungi in their library." After years of research, their scientists reported that they each operated in slightly different ways. To keep them viable, mycelium strains are kept in a freezer set to minus 80 degrees Celsius, or minus 112 degrees Fahrenheit. When ready for production, cultures are brought back up to room temperature and placed in petri dishes along with glycerol and a small amount of protein blend. Later, contents are transferred into flasks, placed in temperature-controlled cabinets, and agitated. After eleven days the mycelia, which now look like floating tapioca pearls, are placed into 3,000-liter fermentation tanks.

Slowly, the mixture is added to progressively bigger tanks—25,000-liter, then 90,000-liter, until as much protein blend as possible has been "treated" by the mycelium.* It's then piped over to a spray dryer, where the process is finished. At this point the mycelium makes up about one percent of the total product. From start to finish the entire process takes about three weeks.

What I saw in Denver was still a bit amorphous. It didn't seem like something I could hold in my hand. It seemed to be floating in jars, hidden inside giant steel tanks, or in poly-lined paper bags. When I finally got ahold of a sample, it was a bar from the Kashi line, which is owned by Kellogg's. If you like to look at labels like I do, and you've seen a simple little line that reads: "pea and rice protein blend," you would have no idea of the complex production that went into that single ingredient. Even stranger is that MycoTechnology, a company built on mycelium, was using fungi solely to process other plant-based ingredients rather than introducing them into our diet.

The amount of refinement our foods undergo is unsettling—and the alternative protein sector is as guilty as any other. In 2020, Myco began delivering its protein blend to JBS, the world's largest processor of fresh beef and pork with annual sales of more than $50 billion. JBS owns many smaller companies, including Planterra Foods, based in Boulder, Colorado. Planterra, which doesn't mention its relationship with JBS on its website, launched a line of plant-based burgers and ground "beef" called Ozo in the summer of 2020. The main ingredient was supplied by Myco, which a company spokesman referred to as "FVP, or fermented vegetable protein."

JBS makes its money on meat, but it isn't going to ignore making more money on the plant-based meat trend. In an interview with Food Navigator, the CEO of Planterra said, "Plant-based isn't going away any time soon." So let's trace this one product to get an idea of

* The end product is the protein blend, which has only a trace amount of the hardworking mycelium.

what it is they're selling us: Peas are grown in North America. Rice, a smaller component of the blend, comes from India and China. Both crops are harvested and shipped by boat to Chuzhou, China, where they're processed into components. Protein is sent back in shipping containers—first by boat to the United States, then by train cargo to Colorado—and processed by MycoTechnology in its giant tanks. After that, it's delivered to Planterra, where it's turned into "meat " and packaged, boxed up, and sent to distribution centers around the country by refrigerated trucks. Finally, after an order is placed, it's stocked in the meat aisle of your local market. This is food, sure, but to me it was only something I'd eat at a lackluster barbecue or, maybe, if I ever dined again at Outback Steakhouse.

We Can Eat Mold?

The leap from mushrooms to mycelium isn't as dramatic as the leap from mycelium to . . . mold. Everyone knows what mold is. We think along the lines of the *Merriam-Webster* definition. Mold is "a superficial often woolly growth produced on damp or decaying organic matter or on living organisms by a fungus." The second definition is less helpful: "Mold is a fungus that produces mold." Kimberlie Le, the twenty-five-year-old CEO of Prime Roots, a Berkeley, California–based startup, prefers not to use the word. She affectionately calls it her "super protein." The "it" here is koji, which is a fungus, a mycelium, and a mold. Even if you've never heard of koji, you've probably already tasted it. In Asia, it's been used for a few thousand years as an active ingredient in seasonings like soy sauce, rice vinegar, miso, and sake.

I met Le at Blue Bottle Coffee in Jack London Square, in Oakland, California. Blue Bottle is an investor in Prime Roots, along with Sweetgreen, companies known more for their traditional foods—lettuce grown in dirt and coffee beans grown in South America—than typical food-tech investors, who ranged from Big Food to Major Ingredients to VCs. (Although Nestlé owns a majority stake in Blue Bottle.) The presence of these investors appeared to be a vote of confidence that what Le and her co-founder, Joshua Nixon, were producing was wholesome food versus lab-made simulacra.

Le and Nixon met in the labs at UC Berkeley. Clichés abound: They loved food, they talked food, and eventually hatched a plan to make food. Next stop was to apply to IndieBio, a synthetic biology accelerator in San Francisco, once they had graduated. Nixon finished with a BS in bioengineering and computer science. Le earned a BS in molecular toxicology, a BA in art, and a minor in music and food systems. Initially, the founders planned to make fish, and joined cohort six of IndieBio under the name Terramino Foods. After they completed the accelerator and raised $4.3 million based on simply their pitch deck, they shifted to highlighting koji instead of trying to make it into something else, which is a common food-science trick. Take surimi, or Krab with a *K*. Glossy and bright white, surimi sells by the boatloads. Made of all manner of fish flesh, but most usually pollock, it's deboned, cleaned, and minced into a paste, then blended with other ingredients. Finally, it's heated and pressed into what kind of looks like a crab leg. This was what Le didn't want to do, but in a sense it was exactly what she was going to do—only instead of making fish, she was making meat.

Le's youth and confidence are charming. Like when she told me that at fifteen years old, she managed "a team" at her parents' food company. In a reversal of sorts, Le's mom, a chef in Vancouver and a celebrity chef in Vietnam, is now the culinary advisor to Prime Roots.

"We feel that beef is well represented, which is why we're focusing on pretty much everything that isn't beef," Le said. She deduced from my barrage of questions that I'd heard her pitch from others and quickly assuaged my concerns: "Everything we're doing is all natural. There's nothing to hide." "Tell me more," I asked. "We're producing a whole new sort of protein," said Le. "Unlike Beyond and Impossible, you name it, those are all ultra-processed foods that are derived from protein isolates, or just separated proteins." Instead, Prime Roots is growing a whole food—koji—and turning it into meat. "It's all done in the kitchen and you don't need extruders,"* she said.

The first time I sampled Prime Roots koji meat, Le and Nixon were dishing it up at a fermentation festival at the Chabot Space and Science Center in Oakland. At tables sprinkled throughout the space were kimchi, koji, and kombucha. Prime Roots served its "meat" in a cabbage cup. It looked like crumbled pork, and tasted like it. As I chewed, I picked up notes of five-spice powder, ginger, garlic, and peppercorn seasonings. I felt like anyone could be happy devouring a dumpling stuffed with it. I went back for several more before I decided I should leave some for everyone else. That night, Le told me they had just signed the lease on a small retail space in South Park, which was once the hub of San Francisco's Internet district.

When I followed up a few months later, it was no surprise to hear that the location had yet to open. A pandemic is nobody's optimal time for a food launch. In spite of the slowdown, Prime Roots continued to improve its koji "bacon" and was selling it on its consumer-facing website. Le said they were working out of a 12,000-square-foot commercial kitchen in Berkeley, but she never let me visit and wouldn't share details despite repeated requests. Last August, when I checked back in, Le shared that Prime Roots closed a series-A investment round for $12 million. This would help them scale their products, and might mean I could finally get my hands on a sample.

"Tell me about the bacon," I asked Le.

"When we make our bacon, we actually form it into a slab—like pork belly—and we put it into a smoker. Then we cut it so that there's an experiential component." Le hoped it would be the uncanny valley† of traditional bacon, meaning it looked so much like bacon we almost can't distinguish it from the real thing. "We don't need all of those things that the other companies use to get texture," noted Le, such as binders,

* An extruder is used to heat and cool ingredients and shape them into finished products like your favorite breakfast cereal. I'll go into more detail on extruders in Chapter 3.

† First coined to describe robots that were humanlike, "uncanny valley" went further to explain that it was appealing until it became too close for comfort.

thickeners, or gelling agents—aka carrageenan, agar, or potato starch. "It comes naturally from koji because it comes out funky and weird already." If by "funky and weird," she meant earthy and savory, I was game.

One of the easier bits of research that I did for this book was eating all the bacon. I tried MorningStar Farms (owned by Kellogg's), Sweet Earth (owned by Nestlé), Lightlife (owned by Maple Leaf Foods), Atlast Foods (a New York startup that was also growing mycelium-based products), and Hooray Foods (a San Francisco Bay Area startup). Both startups had created tasty versions, but certainly not full-stop bacon. They were smoky and crispy, delicious I'd even say, but not fatty enough. When I placed them in a pan, they cooked quickly, which is good if you like it extra crispy, but bad if you're used to slow-cooking, fatty bacon.

This brings up an issue that few startups have addressed well: fat. Most every plant-based company uses coconut oil, which is 90 percent saturated fat, something doctors tell us to limit. Health writer Sophie Egan told me that people think coconut oil is healthy, but it's not. "Any fats that are solid at room temperature are not a good sign." It's not great for sustainability either. Coconut comes from tropical countries that are being damaged to supply our latest food obsessions. Unfortunately, coconut fat is a dream to work with in commercial food formulation. It's the plant-based fat that most closely mimics animal fat. There are a few companies attempting to grow cell-based animal fat such as Modern Meadow, but none were ready to talk.

Before I tried Le's bacon, I purchased a Prime Roots prepared meal at a Whole Foods in San Francisco for $7.99. It had taken so long to get my hands on it that as I stood in line looking over the package, I felt like I was holding a prize. Did the other shoppers know how different this was? The "Vegan Koji Chicken Kung Pao Rice Bowl" was mostly rice, which meant tons of carbs, but sprinkled atop was the koji-based "chicken" I had been hearing about for the past year. It looked like believable little cubes of real chicken, with peanuts and carrots coated in sauce. I cooked it up later that week. The cubes were softer than I expected but held their shape. They didn't have the right stringy texture that chicken has, but that may speak to the need for new language for these new foods. Step one: Let's not call it chicken if it's not chicken.

Months later, my bacon arrived in a cardboard box. As Le described, inside was a slab of bacon with separated, striated strips—just like the real stuff. The only difference was the branding, which was cute and colorful. Definitely not the typical bacon aesthetic. It's hard to not be hopeful about bacon. As directed, I cooked it up with a liberal spoonful of coconut oil, which again: not healthy. Prime Roots' ingredient list was also as Le promised: short and minus any gunk. In the end, the bacon fooled no one. The flavor worked well, but whether crispy or chewy, its texture fell flat. It tasted like either thin, edible particle board (with a smoky taste) or wet cardboard that tasted like I was chewing on a wet paper straw. It was an early version, however, and I continued to be hopeful.

Of all the "future" food being developed, mycelium seems to be the front-runner for improving our food system. It's sustainable, healthy, and it can morph into animal proteins we know (chicken, pork, beef), plus future creations we don't. Chef Dan Barber told me he liked the idea of mycelium. "I'm really interested in it, and would like to learn more," he said, and then hedged: "I'm not against it." In his book *Mycelium Running*, Paul Stamets refers to mycelia as "mycomagicians." And they are magicians capable of building, nourishing, breaking down and disassembling organic molecules. When we take these processes out of the forest floor, do they lose anything or become something different? Let's be sure before we say yes.

After COVID-19 spanned the globe, MycoTech dipped into their library of spores for the benefits many say are inherent in mycelium. Initially, Hahn was making small amounts of supplements for his employees "to be better prepared for COVID-19," he said. "Then we told customers about it and everyone wants to add it to their foods." Soon, four mycelium-based supplements—cordyceps, lion's mane, reishi, and chaga—would be made for commercial sale. Finally, it seemed, the fungi was doing good on its own. I asked Hahn to send me a bottle as soon as he could.

Chapter 3

Pea Protein

Finally, Something That Could Topple Big Soy

The Disneyland of Natural Foods

To understand how sizable the plant-based market has become, I flew to Orange County, California, and attended Expo West. In 2019, the largest natural product show in the United States featured 3,521 companies showcasing their wares to 85,540 attendees. Aisles were packed with humans, all of us sporting a sheen of vitality and wearing head-to-toe athleisure. Samples were handed out liberally—this was almost one year exactly before the COVID-19* pandemic began in earnest—and we chomped, crunched, and slurped our way through the venues without a care in the world.

Warned by a friend who worked in the yogurt industry, I wore sneakers and carried a backpack. As for the deluge of samples, I recalled my "Burger-Off" notes: bite, taste, chew, spit. The show sprawled over a vast

* In 2020, Expo West was canceled just days before it was scheduled to begin because of the COVID-19 pandemic.

swath of real estate, including the entire Anaheim Convention Center, two hotels, *and* their parking lots. Aisle after aisle of vendors displayed a rainbow of innovative foods, enough to torture anyone's waistline. The enthusiasm for plant-based products was intense, even for me, and the reframing of the whole industry, less "vegan activist" and more plant-based "planet-lover," was hugely ambitious.

Vegetarians, meat-free advocates, and flesh avoiders have long been part of our world, but veganism, and the term "vegan," came to the United States by way of England. In 1944, Donald Watson formed The Vegan Society. His passion for not harming animals started early. At fourteen, he proclaimed to his parents that he would no longer eat meat. Slowly, his rules grew to exclude all dairy products. An environmentalist at heart, he grew up to become a woodworker. In order to distance himself from vegetarians—who ate dairy—Watson brought together a small band of ultra-vegetarians to come up with a word that could describe their lifestyle. "Something more concise than 'non-dairy vegetarians,' " they wrote. The group took the first three letters and the last two letters of vegetarian. Donald Watson said it was "the beginning and end" of vegetarianism.

The goal of The Vegan Society was "to seek an end to the use of animals by man for food, commodities, work, hunting, vivisection, and all other uses involving exploitation of animal life by man." This definition illustrates the spectrum of vegans today. At one end: advocates who are fighting for animal rights. At the other: plant-based fans who want better health and a better environment. And in the middle: those of us who sometimes eat bacon. Grabbing on to both of their coattails is Big Food following the trail wherever it goes, hoping for new, high-margin revenue streams and happy shareholders.

The History of Isolates

The idea of abandoning an animal-based diet and turning to plant foods had been simmering for some time. Vegetable proteins were isolated in a laboratory as early as 1930 and were initially used for industrial functions like paper coating. Human food would take another nine years. In 1940, the Glidden Company filed patent number 2,381,407 for an "isolated soybean protein" that could be used in foam retention, and as a stabilizing agent in foods and confections. In 1950, a mocha mix non-dairy creamer made from soy protein isolates hit the market. In 1956, Worthington Foods launched the world's first soy "milk" made from isolated soy protein.

Imported from China since the early nineteenth century for animal feed, soybeans were initially an obscure specialty crop in the United States and Europe. But after the Second World War, the crop gained traction, and revolutionized our food supply. Soy was seen as the answer to the looming "protein crisis" and population boom. Scientists and experts warned of impending food shortages. Newspaper headlines from the time are eerily similar to the ones from today's press that ask how we will feed an estimated 9.8 billion people by 2050.

Widespread use of synthetic fertilizers and chemical pesticides helped double and triple farmers' output. Soy was a low-cost feed for cows that spurred large-scale livestock farming to support the growing middle class. But suddenly we had a glut of grains, and the widespread shortage of food never occurred. In the seventies, the USDA convinced farmers to double down and plant yet more corn and soy as part of the farm support system, which guaranteed farmers income. Alongside government subsidies and promises of a global marketplace for their grain, farmers got in line. This is a compressed snapshot, but it's how we came to the monocrops we have today—wheat, corn, and soy. And it's how soy became the de facto protein in vegetarian foods.

Emerging as the leader in plant-based products—mostly nut loafs and meat extenders—was Worthington Foods. The vegetarian company presented dinner tables laden with recognizable foods, but with futuristic sounding names: Proast, Numete, Tastex, Beta Broth, and Choplets. Sounds delicious, no? Marketing messages promoted "Good Foods for Every Occasion." American food policy during the Second World War assumed that fighting men deserved red meat and that women could make do with protein substitutes. When the war ended in 1945 and meat production resumed, Americans were ready to wash their hands of doing more with less. Over my many conversations with food historian Nadia Berenstein, she pointed out that for most of modern history, "fake" meats (and other "imitation" foods) were low-status, undesirable goods. They were associated either with the deprivations of war or of extreme poverty, or found limited sales among vegetarians and others with dietary restrictions. Once the war was over, Americans wanted to celebrate, and fake meat was no longer the taste of victory.

Worthington's search for novel ingredients led the company to Robert Boyer, a chemist working at Henry Ford's soy research center in Dearborn, Michigan. Ford had grand visions of creating plastic cars out of soybeans. While working for Ford Motor Company, Boyer developed a method for tapping leftover protein extracted from other manufacturing processes and spinning it into fibers. At first his work was aimed at replacing industrial materials in Ford cars including plastics, resins, and lubricants, but eventually he widened his scope to turn the fibers into food.

In the late 1950s, Worthington developed Robert Boyer's spun protein fiber into a line of vegetarian meat. Ralston Purina already had a manufacturing plant for soy, but in 1956, Boyer convinced the company to invest in a food-grade soy-protein-isolate plant. These isolates, higher in protein and with less of that plant-y taste, were spun into fibers using Boyer's license. The company's first commercial launch was FriChik and it was sold in pre-cooked patties.* Other food companies were eager to get in on the new plant-based market. General Mills ramped up its R&D of soy and launched a plant-based meat called Bontrae—Berenstein's guess at its meaning was *bon* for good and *entrée* for dinner.

No company had invested as much in the potential of spun protein as General Mills, and Boyer's methods were at the heart of its synthetic foods

* You can still buy it online today. On Amazon, where it has four stars, one reviewer wrote that it had a "rich gravy and meaty taste."

research program. In the 1960s, General Mills' Isolate Protein Research & Development Program employed more than fifty food scientists hoping to create the next wave of supermarket edibles.

One food scientist told me that while the spun "chicken" was delicious, the manufacturing process was expensive and produced a great deal of wastewater. Despite the marketing spin and attempts to keep its ingredient under wraps—most Americans considered soy as an animal feed—Bontrae failed to take off. General Mills sold its equipment to Dawson Mills, a food processor in Minnesota, and licensed the Bontrae process to Central Soya of Illinois. In the 1980s, both companies abandoned spun protein. Very little, if anything, is made using it now, but Boyer should certainly be considered one of the grandfathers of all things meat-like.

Ultimately, Worthington Foods refocused its attention on a cheaper formulation by using textured vegetable protein (TVP), which then was made from soy flakes. In 1975, it launched the now-familiar MorningStar Farms brand to the nation. The soy-based line of faux meats was carried in supermarkets and grocery stores across the country. As the blueprint for today's plant-based startups, Worthington Foods became the largest US company hoping to convince the country to eat its vegetables.

Today, Worthington and MorningStar Farms, which was the top seller of plant-based food products in the United States in 2019, are both owned by Kellogg's. While Impossible and Beyond steal today's media chatter, these two vintage Kellogg's brands already have widespread distribution to support our daily diet including plant-based burgers, breakfast sausages, and chicken tenders. Today's "wholesome" plant-based market is quickly becoming oversaturated with options from every Big Food producer posing their products as the healthiest, highest-protein, and most delicious. If we look under the hood, it's likely that they all share ingredient suppliers, formulation recipes, and co-manufacturers. What's especially interesting to note is that soy began as a technology platform for creating other foods. This same approach is being used with peas.

Making Meat Out of Powder

Textured vegetable protein sounds unappetizing. It's a processed, industrial-grade food, but it isn't necessarily unhealthy. Invented in the 1960s, it's something of a lynchpin for helping meat analogues get closer in texture and appearance to traditional meat. To create TVP, whole soybeans are processed to isolate the protein, removing the fiber and starch. Wet at this point, the protein is spray-dried and then put through a high-heat extrusion machine, which has been a standard in food processing over the past fifty years. Extruders are used in making hundreds of processed foods. In the early days of extrusion, the machine mostly spit out macaroni and cereal pellets, but by the 1980s it was a fast-working, high-temperature bioreactor that could transform raw ingredients into ready-to-eat finished products like croutons, crackers, and, yes, baby food.*

TVP looks like small, inconsistently shaped pieces of Cap'n Crunch cereal without the berries. You could toss the crunchy, flavorless pellets in a bowl, but why would you? Unless you happen to be Ethan Brown. True story: In his early Beyond Meat days, the CEO and founder would grab handfuls of pea protein puffs from sample bags at the office, soak them in plant-based milk, and eat them for breakfast. Yum. You can't question his dedication.

Early iterations of TVP were made from isolates because the higher protein content allowed for better adhesion—no one wants their burger to fall apart. One of the first people to get to a convincing texture of chicken using TVP was Fu-hung Hsieh. Born in Taiwan, and educated in the United States, Hsieh is a slender man with a gentle voice. He told me he was mostly vegan but occasionally ate meat. *Bacon?* I wondered. A biological engineer and food scientist by training, Hsieh spent his career tinkering with food—the opposite of a farmer, but with goals that

* Because of chemical changes during processing, some manufacturers spray vitamin solutions onto their products after extrusion to correct perceived nutrient deficiencies.

were similar: feeding the world, the earlier version of making the world a better place.

One of his first jobs was at Quaker Oats, where Hsieh holds four patents on food formulation "enhancements." (He used glycerin to keep raisins soft and upped the fiber in oat bran by adding beta-glucan.) After earning a doctorate in food science in 1975 at the University of Minnesota, Hsieh joined the faculty at the University of Missouri. During our phone call, he recalled the early attempts at veggie burgers that McDonald's offered in the 1990s (and again in the early 2000s). "The taste was terrible," he said. One of the ingredients was an early rendition of TVP. "It just was not meat-like. It didn't have meat texture, appearance, or bite," he said. With the luxury of a university job that allowed him flexibility and time, Hsieh worked on making a better version. "If we could make something look like real meat, then we could get the consumer to try it," he told me.

It took Hsieh more than a decade to perfect his chicken analogue, working with a team of graduate students; his colleague, Harold Huff; plus a tractor-shaped machine called the APV Baker 50 mm twin-screw co-rotating extruder. "We were fortunate we had a very industrial, pilot-size extruder," Hsieh said. Initially, they used various isolates including soy, pea, and whey. Acceptance of uncommon protein has come so far that Hsieh laughed, and said, "Insect protein could be used." I haven't seen it in a burger, but it's already in paleo energy bars.

The APV Baker is a hulking steel contraption that looks somewhere between a tractor and a massive Xerox copy machine. When ingredients are forced through the extruder's small opening, they're cooked inside the unit with high-pressure shear forces, and heat created by the screws encased in the barrel. When spit out, materials often puff up due to the release of pressure and conversion of water into steam. You can watch low-quality YouTube videos of people standing on ladders tossing ingredients into the top end of an extruder, while finished items shoot out the other end. TikTok videos probably exist, or they will soon.

To feed the world, food manufacturing has replaced nature. A cow raised for meat takes about nine months from birth to burger. By contrast, an extruder can turn plants into chicken in about one minute. "It blends, kneads, cooks, cools, and forms [the ingredients] in one continuous motion," Huff told the University of Missouri alumni magazine. Hsieh and Huff patented their process for "chicken" in 2011.

After Ethan Brown read about the patent, he licensed the technology. In 2012, chicken-free strips were Beyond Meat's first launch in Whole Foods Markets in Northern California. As part of his agreement, Brown was required to build a manufacturing plant in Columbia, Missouri. Today, it's one of several plants Beyond Meat has pumping out fake meat. Like many public companies, Beyond Meat has hurdles to overcome: It's dealing with litigation from a former co-manufacturer over unpaid invoices, and some he-said-she-said over trade secrets; and the El Segundo company had its burger dropped from McDonald's menu trials.

In January 2020, McDonald's began testing what it called its "PLT" using a Beyond Meat burger patty at twenty-four locations in Canada, but in April the fast-food chain ended the trial and offered no plans for bringing it back. Shares of Beyond stock dropped 7 percent the day after McDonald's made its announcement.* Tim Hortons, a Canadian chain with 4,800 locations in North America, pulled all of its Beyond products by January 2020, because, a company spokesperson told Reuters, "The product was not embraced by our guests as we thought it would be." And Beyond Meat's first product, the chicken strips made with Hsieh's patent, have been pulled off the shelves, because, per Beyond Meat, the strips "weren't delivering the same plant-based meat experience" as its other meat analogues. On a positive note, Kentucky Fried Chicken has been

* Based on learnings from its PLT trials, McDonald's announced in November 2020 that it would launch "McPlant," its very own line of plant-based burgers, in global markets the next year. Initially, the fast-food chain distanced itself from Beyond Meat, but a few months later the two companies announced they would be working exclusively with each other to create the McPlant line.

running small-scale tests of a Beyond Meat "fried chicken" at locations in Southern California. It's rumored to be delicious.

I talked over our insatiable desire for animal meat with Dr. Deborah A. Cohen, a senior scientist at the RAND Corporation, and author of *A Big Fat Crisis: The Hidden Forces Behind the Obesity Epidemic—and How We Can End It* (2013). "There are a lot of myths about diet," she said. "It's just habits that people say they need meat. We already have more protein than we need in America. We don't have anyone with protein malnutrition." She mentioned places in Africa that lack access to life-sustaining protein, countries we frequently overlook when we talk about access and need. In fact-checking her statement about the United States, I learned that 98 percent of the country gets more protein than the daily recommended amount, yet it's the one thing everyone is most concerned about getting. Way back in 1971, in her book *Diet for a Small Planet,* Frances Moore Lappé noted that "most Americans consume twice the protein their bodies can use." An unchanged theme that shows no signs of stopping.

"Plants have more nutrients," Cohen continued. "It depends on how much it's processed. You could have fruits and vegetables and grains where all the nutrients are taken out of them." One example of this is a puffed green pea snack. One serving of the processed version has more than twice the calories, more than five times the fat, and 1.5 times the carbohydrates. Processed, the pea simulacrum loses most of its vitamins and phytonutrients—components that are hard to consider line by line but that promote better health. To be beneficial to our diet, foods "need to have their nutrients," said Cohen. Or, simply put: Our food still needs to be food.

This same thing can be seen in whole grain bread. "Whole grain" refers to a wheat-flour product made up of the bran, endosperm, and germ from a grain of wheat. Normally, making bread is a simple process. Wheat is milled into flour and baked into bread. In the industrial version, however—what's available at most supermarkets—the bran, endosperm, and germ are purchased as processed ingredients from other manufacturers, then mixed back together and baked. Cohen noted, "It's not a number of ingredients, it's chemicals. It's not processing, it's refining that's the problem." Whether we call it processed or refined, what's important to note is that many of these foods being marketed to us as "healthier" are not. Because they've been churned through our industrialized food system, we've lost some of the original nutritional content. Companies promote their products as "healthy," but they are accountable for their profits, not for the actual health of their customers. There is no incentive for products to be as healthy as possible.

When the World Turned to Legumes

The second time I met Adam Lowry, the CEO of Ripple Foods, it was through a haze of déjà vu. We were in the same conference room, in the same Manhattan high-rise, at the same fancy public relations firm, and Lowry was presenting another non-dairy creation. The first time we met, I tried Ripple's plant-based milk, which is made from peas. The headline I imagined, and which ended up a reality, was a bit jokey: *Are you prepared to drink pea milk?* It felt like a shopping list for my bomb shelter, and sold with my kitchen spirulina maker. My personal response, while cliché, was true: That pea milk was a slam dunk. When it launched at Whole Foods in 2016, I bought it. I still buy it today when I see it in the dairy case. The protein is nice, but I especially like the thick and creamy mouthfeel.

Peas have an aura of wholesomeness, one we associate with "pure" baby foods and children being told to "eat your veggies." While that's the image of fresh green peas, the peas being used by Ripple are the ones you might throw into dal or split pea soup. These are field peas, *Pisum sativum*, a boon to farmers for crop rotation—they fix nitrogen in the earth and are drought tolerant. They're good for us, too. We *will* actually be healthier if we eat them. One serving of peas—half a cup—has almost 9 grams of fiber, a nutrient that is lacking in the diet of most Americans.

While I quickly put Ripple's pea milk in my rotation, the second product I tried—pea yogurt—seemed iffy. Spread across the conference room table in New York were several plastic containers with no branding. I glanced down at the dull colors, and with a new appreciation for toddlers everywhere, I make an "icky" face in my mind. To develop the yogurt, Ripple hired a flavor house—a common practice for most food companies. With its help, Ripple created what are usually crowd pleasers—blueberry, strawberry, vanilla, and so on. I gamely dipped a tiny plastic spoon into each sample. The real test was the plain yogurt, my go-to because it has less added sugar. Putty-gray with a thin consistency, the plain had an unpleasant, vegetal taste. Without fruit to mask the off-flavors, the plain was fairly unpleasant. Using peas in any novel application beyond rolling around on our plate is a challenge that requires extensive processing. Off-flavors come primarily from the compounds in the skin that give peas their color, and also a class of compounds called phenolic acids, which are generally associated with sour, bitter, and astringent flavor attributes within plant protein isolates. Despite the valiant effort, Ripple's pea yogurt was not destined for my fridge.

In addition to his entrepreneurial verve, Lowry is a competitive sailor. His first company, Method, was a redux of the bad-for-you cleaning products, only reformulated without toxic chemicals and designed so that we didn't need to hide it under the counter. Lowry built it into a $100 million company before selling it to Ecover, a Belgian cleaning company, in 2013. Lowry and his co-founder, Neil Renninger, met on the speakers circuit—technology conferences and business summits that accompany the founding of a successful company. Renninger formed and sold Amyris, a biopharmaceutical drug discovery company in the San Francisco Bay Area.

In 2014, when Lowry and Renninger put their heads together to contemplate food categories that needed shaking up, dairy was at the top. Options existed but most lacked any significant amount of protein, which was newly becoming a breakout category in the supermarket. The challenge for Ripple Foods was simple: Break down the known (milk from a cow) into key components, vitamins and minerals and protein that could be repackaged into a better-for-you version, with none of the animal welfare or environmental baggage. Consumer mindset had begun to shift, and protein callouts on the front of the package allowed Ripple to stand out. "Highlighting the presence of protein in plants serves to reorient culturally conditioned views of animal products as the best or only source of this nutrient," wrote University of Oxford researcher Dr. Alexandra Sexton in "Framing the Future of Food," a 2018 article about the contested promises of alternative proteins.

With Renninger as the scientist and Lowry as the businessman, Ripple worked to innovate at the ingredient level. "That's the real hallmark for the food industry," Renninger told me in 2016. Similar to the ways in which faux-meat makers are processing plant material for their unique needs, Ripple rooted around in the plant kingdom for commonly grown plants with a high protein content that could stand in for whey, one of two key proteins in cow's milk. "I think the entire space would agree that there's potential impact by focusing on the ingredient," said Renninger. Here we see again the changeover from something simple and edible in its own right to a technology tapped for its functional benefits.

Homing in on legumes, the Ripple team tried out various beans, including lentils and soy, navy, white, and mung beans. "The process works on all of them, some better than others," said Renninger. Ultimately, they chose peas because it was a cheap crop, and there was an existing supply chain for yellow peas.

Late in 2019, I met up with Lowry again. I inquired about that yogurt I had tasted. He joked that it had all of 1.5 fans. "We messed up on the product," he told me. The problem was that it was gritty, he said, referring to their proprietary pea protein blend. "We weren't satisfied and we discontinued it." Ripple's headquarters are in Berkeley, California, in the former Pyramid Brewing location. We sat in a restaurant-style booth and through the giant picture windows next to our table I could see into the lab, which used to be the brewery. I noted it was quiet, and Lowry pointed out it was Friday.

Ripple thinks it's different from other companies because it purportedly created the purest version of protein from peas, with the cleanest taste. I had the chance to try a version of pea-based milk from MycoTechnology—another company that claims it has created a better-tasting pea protein. The two were a close match. They still tasted like peas. As companies enter the plant-based sector and look for funding, they need a reason to secure investment dollars. Founders tell investors one of two things: They have proprietary technology* or they have the cleanest, purest, best flavor. First, it's a fight for funding from investors. Then, it's a battle for the wallets of consumers.

One would think that creating the cleanest protein with a low vegetable aftertaste would be extremely valuable. I've asked Lowry several times if he planned to sell Ripple's protein isolate to other companies, and he always said no. I assumed his main reason was because it was like their "Intel inside" platform. They could license their technology if they wanted to, but so far they have not. Impossible doesn't sell its heme either. These are ways to set their brands apart and line up consumers. Lowry tells me he's still working to lower the price of Ripple's milk, which speaks to an inability to make the protein cheap enough for other food manufacturers.

After the interview, Lowry took me to the staff kitchen and had me taste the new yogurts. They still had that same taupe-y gray hue that seemed to follow pea protein wherever it went. While the yogurts were better, they weren't better by much. In 2020, I wasn't seeing this yogurt widely at the supermarket. Lowry wrote over email that it was because of the pandemic and grocery store limitations on picking up new products. Ripple had also launched ice cream, which seemed an easier sell, at least to me, but I couldn't find that either.

If the protein in my diet were only coming from isolates or concentrates in processed foods, would it be missing key nutrients? To get a better handle on this dilemma, I turned to Dr. Michael Greger, a clinical doctor and author of the books *How Not to Die* and *How Not to Diet*. These titles may sound alarms, but Dr. Greger has a way of making the subject feel less threatening. When I saw him onstage at the Plant-Based World expo in Manhattan in 2018, he paced back and forth quickly, his

* Most investors have told me they look for a companies with unique technology (not deliciousness) as one of their primary requirements for funding.

black suit falling loosely around his vegan-lean frame. Visuals behind his head were created to shock, and his devotion for what he was saying was hard to resist. "Food is a zero-sum game," he told me. "When we stick something in our mouths, we have lost the opportunity to stick something else in." I laughed, then asked whether protein isolates were good for me, or at least better than meat.

"From a nutritional [angle] it doesn't make sense. Just like isolated fat is bad, isolated carbs are bad, refined protein is bad. Bad meaning you are stripping out the macronutrients. You have removed all the nutrition." In the same way that we lose nutritional components such as fiber when we split something up, Dr. Greger pointed to another loss from whole foods. "There are phytonutrients that have yet to be named before that aren't on a label," he said. Phytonutrients (*phyto* means "plants" in Greek) are (possibly) disease-preventing nutrients found in whole foods. According to Dr. Greger's website, NutritionFacts.org, fruits and vegetables from nature contain more than 100,000 different nutrients, but you won't find them in plants from New Food companies that have been fractionated into components. "When you think of legumes, they have all the things you want that animal proteins have, but minus the bad stuff," he told me. The bad stuff: saturated fat, additives, and chemicals.

Once processed into solely protein, legumes and beans no longer exist, and have lost much of their benefits. If you were to eat whole peas, you would get phytonutrients like carotenoid, lutein, and zeaxanthin, which are said to promote vision and eye health. Whether it's pea protein in a burger or a pill: "When it comes to food, the whole is often greater than the sum of its parts," said Dr. Greger. But as a clinical doctor promoting plant-based diets, he was happy to see this as the first step toward shifting to plants—even if it *was* a glass of milk made of pea protein isolates.

Today's "It" Ingredient

Peas would be nowhere if not for the fickle American diet, and our on-again off-again love affair with soy. The soybean crop has its merits. It's cheap, plentiful, and easy to grow. Like peas, it also fixes nitrogen in the soil, a chemical element that is a natural fertilizer and allows for easy crop rotation. Soybeans feed our animals, which feed us. For humans, soy includes the nine essential amino acids that our bodies need to function properly. Our bodies actually do create amino acids, but not the essential amino acids that support normal bodily functions; and we don't store excess amounts, which means we must consume them daily in food. Because of this, soy is a mainstay of almost every plant-based food product on the supermarket shelf.

Amino acids are at the heart of a raging battle over the merits of getting our nutritional needs met from animal protein versus plants. A clear head in this area is Christopher Gardner, a scientist at Stanford Prevention Research Center. His interests have led him to run many nutritional studies funded by the National Institutes of Health (NIH). More recently his focus has been on what he calls "stealth nutrition," or ways that public health professionals can use non-health-related strategies to improve the human condition. An example of this is removing the ubiquitous trays in university cafeterias—this one shift prompted students to eat less food.

I first met Gardner at a conference at the Culinary Institute of America in Hyde Park, New York. He was very approachable with a relaxed, California vibe—he wore Birkenstocks and his painted toenails peeked through the ends. We talked over his research. In a protein study he co-authored in 2019, he wrote that a typical day of eating a variety of common foods would include adequate amounts of both essential and nonessential amino acids, almost regardless of the presence of animal foods. On the topic of a solely plant-based diet, Gardner assured me that in most ways it was superior to animal-based. "Animal foods don't have fiber, and plant foods generally don't have saturated fat. It's actually more beneficial to choose plant foods over animal foods," he said.

In August 2020, Stanford released findings from a trial run by Gardner on the effects of eating plant-based "meat" compared with animal-based meat on levels of trimethylamine-N-oxide (TMAO), a compound that some researchers say elevates cardiovascular disease risk factors. For eight weeks, thirty-six participants were asked to consume two servings a day of either Beyond Meat or traditional animal meat. (In addition to their usual diet.) After eight weeks, they switched to the other protein source. In Gardner's study it was found that participants who ate traditional meat first, followed by Beyond Meat, which is primarily pea protein, showed lower LDL, body weight, and TMAO. No lowering was shown if it was the other way around, however. Now, having met Gardner, I read over this study with much interest, and didn't miss spying that Beyond Meat funded the study and supplied product to the participants. Marion Nestle has a book on this very topic: *Unsavory Truth: How Food Companies Skew the Science of What We Eat* (2018). Nestle also thinks highly of Gardner, but she wondered in her "Food Politics" newsletter what else the participants were eating and the significance of the TMAO measurements. These are great questions. Extrapolating from this study, Nestle wrote that "two servings a day of Beyond Meat is unlikely to be harmful, but whether substituting Beyond Meat for real meat is truly useful for health in the absence of other dietary changes remains to be confirmed, hopefully by independently funded research." I'd like to see this too.

We are becoming more aware of our individual needs as humans. And many are choosing to burrow deeper into the nutritional supports necessary to fuel our ambitions and desires—build muscle, lose fat, improve immunity. In response, today's food makers are rapidly formulating products to satisfy any number of specialty diets out there. As our dietary goals diverge further, manufacturers are churning out snacks du jour for Whole30, keto, paleo, low-FODMAP, caveman, blood type, and OMAD. Want a seasonal probiotic? You can get that. Want to eat according to your own personal microbiome? You can do that! Send your poop out to be diagnosed and then eat your meals based on what's found in your stool.

Back to soy, which is still widely used because it's cheap and abundant. But it has dings against it. Some of the proteins found in soy make it one of the top eight major food allergens. As a crop, it's predominantly planted using genetically modified (GM) seeds. And it has long been maligned as cancer causing, which numerous clinical studies refute, and many point to consuming soy—like tofu—as healthful. However, because of our wide range of differences in processing, digestion, and ethnicity, making blanket statements about soy is exceedingly complicated. For these reasons, today's marketplace has frequently abandoned soy in search of better, newer, sexier forms of protein. Peas are in the limelight today, in part because they are mostly considered non-allergenic;* and as a crop, peas are not GM. But there's a host of other contenders waiting to take pea's spot, including mung bean, fava, and canola.

When I talk about peas here, I don't mean the whole vegetable, but rather the protein that is extracted from the legume. On a label it might read "pea protein isolate, concentrate, flour, or protein." What makes a protein an isolate is its purity. It must be more than 90 percent protein to be labeled an isolate. At 70 percent, it's a concentrate. "Pea protein isolate" isn't one of those phrases that rolls off your tongue. And it shouldn't. Isolates require extensive processing.

Peas are grown, picked when dried, and shipped to a manufacturing plant where the molecules can be split apart into protein, fiber, and starch. This process, which can be done wet or dry, is called fractionation. Many of these manufacturing facilities are located in China, but the majority of the crop comes from North America. Chinese manufacturers keep the starch, which they turn into noodles, and ship the protein back to the United States. Turning a pea into three buckets of goods makes the ingredient far more valuable. Pea protein as a powder, versus the whole legume, is also much easier to make into other foods. The downside is that shipping peas back and forth to China extends the carbon footprint and creates a global supply chain that is affected by tariffs and disease outbreaks. Alan Hahn of MycoTechnology told me that buying his ingredients from China was still cheaper, despite the tariffs, than buying from

* But peas are in the legume family, as are soy, mung bean, and fava (but not canola). This means they're closely related to peanuts. Despite this, few studies exist that have investigated the allergenicity from peas in humans.

the United States. When the COVID-19 pandemic hit, the supply chain slowed down by a few weeks, but Hahn had only good things to say about his Chinese partner. "They have been really nice to us. They sent us cases of [face] masks."

Ripple's process stays in the United States. It begins with ground, dried peas grown in North America. Pea powder is soaked to make it soluble. Proteins are extracted using a combination of temperature, salt, and pH—a measure of the acidity or alkalinity of a solution. Then the flavor, color, and carbohydrate molecules are separated to retrieve the protein from the liquid. To retrieve the protein base, Ripple uses a centrifuge—a machine with a rotating core that can separate liquids from solids—to spin the mass. Once the protein concentrate is separated from the fat, fiber, and starch, the wet paste is turned into milk. "From beginning to end the process takes about two hours," Renninger said.

A dairy cow is no match for Ripple's two-hour procedure. In twenty-four hours a cow can produce six to seven gallons of milk, but it doesn't spring from an udder without human intervention. Cows must be pregnant every year to produce milk. To do this, dairy cows are artificially inseminated to keep their flow of milk going. Our industrial supply of milk, and cows being artificially inseminated to keep producing, continues because milk industry advocates have worked for decades to make cow's milk a staple of human life. Humans are programmed to drink our mothers' milk, which is high in fat and carbohydrates, but actually low in protein.

Cow's milk is also processed milk. What cows eat, where they graze, how they're wintered, whether their barn is clean—these are all factors that impact the final product. Once milk is obtained, it's refrigerated, and then pasteurized through heating. Processing happens on the farm, and in the creamery, just as it does in the lab. The question here is whether a healthy diet is better if it includes cow's milk, almond milk, oat milk, or pea protein isolate.

Using plants as components is nothing new—where would our sandwiches be without bread? But after decades of monocropping, today's food manufacturers are finally looking beyond modern staple foods, and paging through a growing database that is the universe of plants. Out of 391,000 possible species, the edible number ranges from 7,000 to 30,000. Yet, agronomists tell us that while we've tried some 3,000 species of plants, fewer than 200 have become staples in our diet.

One company working to dramatically increase the number of plants that we see on our plate is Eat Just. The San Francisco–based company is known to some for its legal problems (calling its plant-based "mayonnaise" Just Mayo, which might lead shoppers to think the product is actual mayonnaise made from eggs) and its business problems (buying back that mayo to increase its sales figures) than for its knowledge of plants.

Just claims to have analyzed thousands of plant samples from more than seventy countries in order to identify novel proteins. Before you imagine a library you can walk into and browse for gardening ideas, remember that whatever the company has discovered is carefully guarded intellectual property. Its intellectual property—more than fifty patents have been filed (and some have been bought from other companies) to date. This has helped the company raise more than $220 million and retain its billion-dollar valuation with investors. The potential for food manufacturers is huge. If you can be the first to commercialize a novel ingredient, future success on Wall Street is almost a certainty. It's the commodification of nature itself.

To create shelf-stable mayonnaise, Just scientists tested more than one thousand formulations of the recipe to find the best proteins for foaming, gelling, and moisture retention. In spite of those tests, the plant they selected was the well-known and widely used canola seed. For its next commercial launch, Just used mung bean isolates to make a liquid egg. Mung beans, like the pea, are in the legume family. It took the company four years to refine the recipe and bring the product to market. Now that it has consumer acceptance in the United States, the product is making its way to Asia, including Hong Kong, South Korea, and China.

The US government is also working to further our knowledge of these plants. As part of a multi-year program called the Pulse Crop Health Initiative (PCHI), the USDA is funding many projects that will increase our knowledge of pulses, the word for the dried bean versus fresh, and including most (but not all) legumes. Rebecca McGee, a plant breeder in the Agricultural Research Service (ARS), a division of the USDA, has a grant from PCHI to help understand peas better. McGee had what I've come to call "plant-breeder humor"—dry and poignant, but only when it concerned plants. "I always tell people that my primary objective as a plant breeder is yield. If other breeders say their objective is any different, they're lying," she said.

According to McGee, peas have on average 22 percent protein. But "there's a significant amount of genetic variation in the protein content." McGee's current project is to find the genetic nature of this fluctuation in pea protein levels. PCHI funded McGee's project, called MP3—cutely named for its research goal of driving more peas, more protein, and more profit.

"We scoured the world looking for yellow peas," McGee said. She anticipates "looking at five hundred yellow pea lines." Recently, McGee collaborated on sequencing the entire pea genome with a team of scientists around the globe. "The pea genome is huge," she said. It's more than 1.4 times larger than our human genome (4.5 gigabases—a unit of length for DNA molecules—against about 3.2). But "it's full of junk," she said, with many highly repetitive sequences.

While we were on the phone, I had to ask: Why do legumes make me so gassy? She said it was because they contained low-molecular-weight carbohydrates. "Uhhh, and that means?" I asked. Apparently, it just means small. "We scientists always like to make things more complicated," she said with a laugh. Wanting to know more, but not wanting to use up the last few minutes left in our call, I read up on pulses in the textbook *Foods: Facts and Principles*.

That bloated and gassy feeling occurs because legumes include oligosaccharides, which are sugars from the raffinose family that are "notorious for flatulence production." Because we're missing a key enzyme needed to digest them (α-galactosidase), these sugars escape digestion when we dine on them. Oligosaccharides aren't absorbed and, instead, are digested by the microflora in our lower intestinal tract. Have your eyes glazed over yet? As a result, legume-loving humans produce large amounts of carbon dioxide and hydrogen, and a small amount of methane.

"Beans are the most concentrated source of fiber and minerals," Dr. Greger told me. "The only place fiber is found in abundance is unprocessed foods." They're just as good for our soil. In a European study from 2016 on the benefits of introducing legumes into crop rotations, researchers found that diversifying crops showed a 20 to 30 percent drop in nitrous oxide emissions, with a 25 to 40 percent drop in fertilizer use. More important, it also showed that "positive environmental effects do not necessarily mean gross margins go down." Planting a wide variety of crops may be the most important thing we can do to slow climate change. Peas, mung beans, chickpeas, and every other nameless pulse, bean, or legume that we haven't tested widely yet can help restore our soil and improve our diets. Far from some fancy tech startup with millions of dollars in funding, this basic farming solution could mean we wouldn't have to choose between our health and the environment.

One company looking past the soy and pea starting bench is Climax Foods, based in Berkeley, California, and founded by Oliver Zahn, an astrophysicist who's worked for SpaceX and Google. Climax raised $7.5 million in an "oversubscribed seed round" with little to show other than Zahn's résumé and his promise to investors that he would use unorthodox solutions to improve what he called "suboptimally created foods." We talked on the phone during the COVID-19 pandemic, each of us safe in our own homes. Zahn assured me that the number of ways you can assemble plant products to make animal products is so large that unless you have predictive models and data screening algorithms, there would be no way to tackle it. "This is better for the environment, and a much safer product to eat," he claimed. In light of COVID-19's hop from wild animals to humans, it was hard not to disagree that plant products were safer. Even the shelf life of plant-based foods is longer than anything else, a fact that translated to fewer trips to the market, which was better for shoppers, workers, and vulnerable populations.

In 2019, when I grazed my way through the aisles of ExpoWest, I spotted more flavor innovation than ingredient diversity. Outside of the pili nut, little there surprised me. The dilemma for startups is that they may initially launch using lesser-known ingredients, but eventually there's a shortfalling when they attempt to scale it up to mass consumer levels. At this point, cheap and available becomes the only choice. This is why MycoTechnology went to China for its peas and why peas won out in Ripple's milk. But peas aren't perfect. Especially in yogurt!

When we spy pea protein being touted on a package, there's an illusion that we know what it is. We imagine a field of lush, verdant peas. But this is a mirage our brains insert when reading the label. Industrial food, or ultra-processed food, is very different from anything we can make in our kitchen, and the end products we're buying have been modified and treated in so many ways that our simplistic understanding of what food is will fail to match up with what we're actually ingesting. "If we really wanted nutrition, we would center our diet around whole foods," said Dr. Greger. "The intake of legumes—beans, split peas, chickpeas, and lentils—appears to be the most important dietary predictor of survival in older people around the globe, with an 8 percent reduction in risk of premature death for every one-ounce increase in daily intake." It seems clear that peas will stick closely to the path of Big Soy, becoming the next wholesome plant turned into technological system. In twenty years, will this "pea technology" be considered a good thing for humans?

Chapter 4

Milk and Eggs

If No Animals Were Involved, Is It Vegan?

Cowless Milk

I generally avoid Forty-Second Street with every cell in my body, but the hotel I was going to was smack in the middle of Times Square in New York City. When I exited the subway, I darted for revolving doors that led me into the sanity of the hotel lobby. I took an escalator up to the second floor and found myself in the center of a small welcome area. Silver urns of coffee, plates of cut melon, and bowls of yogurt and granola were set out. The crowd was mostly male, mostly white, and they moved around the room like bumper cars. I looked for a familiar face.

It was 2016, and it was the very first installment of Future Food Tech. By 2019 the event welcomed thousands of attendees, but in 2016 it had yet to become the staple it is today in promoting the plant-and cell-based industry of future foods; there were just two hundred of us.

The agenda for the conference leaned heavily on science, and over the next two days I was hoping to take it all in like osmosis. One company I wanted to learn more about was Muufri, whether because of the goofy name, or because of its mission. Whereas most everyone in food-tech was focused on replicating animal meat, Muufri was re-creating milk. The co-founders, Ryan Pandya and Perumal Gandhi, who are vegan, had told me they missed one thing above all else: cheese.

Stroll by the dairy aisle in your local market, and chances are good you'll find an impressive array of vegan cheeses. Should you live near a Wegmans, go there. According to a 2020 report released by the Good Food Institute and the Plant Based Foods Association, the East Coast supermarket chain has the most vegan products stocked on the shelves of any grocery store chain. Overall, these vegan cheeses are having no problem selling. The herbed "cheese" spreads are actually quite good, and the sliced cheeses (sometimes) melt. By far the best match in taste and texture is cream cheese, but there aren't any soft, gooey washed-rind cheeses like a Camembert or a Brie; there aren't any cheeses that melt convincingly in the oven like mozzarella—the most consumed cheese in America.

The vegans at Future Food Tech were easy to spot. Dressed in loose jeans and casual shirts, they were younger than everyone else. Gandhi was wearing one of those generic black SwissGear backpacks. We hadn't met in person yet, so I walked up and introduced myself. We made small talk until Pandya pointed his thumb toward the backpack: "We have samples. Don't tell anyone," he said. I widened my eyes. "Mmmkay," I said. "Do I get to try it?" The pair looked around and motioned for me to follow them. I felt like I was buying a fake designer bag on Canal Street. We walked down a long hallway that veered away from the main room.

Pandya unzipped the backpack and pulled out a glass bottle with a whitish-yellow liquid inside. Gandhi, looking furtively over his shoulder, explained that they didn't have enough to share with everyone at the conference. The bottle was only half full. He poured a tiny amount into a plastic cup. Not wanting to gulp it down, I delicately poured the liquid into my mouth. My first thought was that it was thin—maybe too watery. Then, too sweet. I took another sip. I wasn't sure if it was supposed to match whole milk or nonfat—what I grew up drinking. And never liked. My face must have betrayed my thoughts. "We're still working on it," they both said, then stashed the bottle in their backpack and we rejoined the other attendees.

Two years prior, in 2014, and with only the barest of ideas—to make milk proteins using microbial fermentation—Pandya and Gandhi applied to a synthetic biology accelerator in Cork, Ireland. They were accepted and given $30,000, research space, and mentorship to get their idea off the ground. When Pandya's mom heard the news, she said he should be careful about giving money to strangers. When she learned that it was the other way around, she laughed with surprise.

Eventually, that accelerator—since named IndieBio—has become the preeminent SF Bay Area network for fostering advances in synthetic biology. Over the next three years, I talked to Pandya and Gandhi often. Each time they raised more money they sent me an email, perhaps hoping I would write a story. Slowly their post-collegiate scruff was replaced with a more formal polish. In 2016, they dropped their final bit of quirkiness, the company's name. Muufri ("moo-free," get it?) became Perfect Day, a name that was indirectly inspired by a study done by another pair of scientists who wanted to track the happiness of cows that listened to music. According to the founders, Lou Reed—who penned a song titled "Perfect Day"—was the bomb among bovines.

The History of Microbial Fermentation

Many of the future foods in my book are here because of people like me. Or rather because scientists had successfully isolated human insulin, which is purchased by people like me, those with type 1 diabetes.

Insulin was discovered in 1889 by a pair of German researchers who performed some unusual (and many would say unethical) tests on dogs—they removed Fido's pancreas, and the dog presented with diabetes-like symptoms. When they injected pancreatic juice back in, the dog's condition improved. The theory was that this lifesaving protein was located in a very specific cluster of cells deep within the pancreas.

This cluster was eventually named the islets of Langerhans, after the researcher who found the cluster, but not the Germans who removed the dog's pancreas. Because insulin was found in the islets of Langerhans (as a teenager I always pronounced it as "Langerhorns," like they were a football team from Texas), researchers settled on the Latin word *insula,* meaning "island," for the " isles" that held this lifesaving medicine.

In 1922, Frederick Banting, a Canadian surgeon, was the first to try insulin on a human. It worked. The next year, Eli Lilly and Company began to mass-produce insulin, not from dogs, but still from dead animals—pigs and cows. This helped, but it often couldn't meet demand, and the potency varied. I was twelve years old when I was diagnosed with type 1 diabetes in 1983. After I successfully demonstrated I could handle a needle by giving an insulin shot to an orange, the nurses handed me my own bottle to take home. The tiny vial held Humulin insulin and stamped in the corner in cursive was a name I would come to know very well: Eli Lilly. Underneath the word "insulin" it read: recombinant DNA origin.

Approved by the Food and Drug Administration in 1982, less than one year before my diagnosis, Humulin was the first recombinant pharmaceutical approved for use in the United States. The technology, discovered by Genentech, and licensed to Lilly, allowed scientists to insert a human gene coding for the insulin protein into a common bacterium (*E. coli* in this example), turning it into a temporary host. Under the right conditions, this bacterium could generate a biosimilar human insulin. The host (the *E. coli*) and the insulin (the protein) could be separated in a purification process, becoming purer than what could be removed from a live human pancreas.

In the food industry this process of tweaking genes to generate a desired output is referred to in shorthand as "microbial fermentation." We don't have to look far to find examples of it in the food we eat. One of the ingredients in cheesemaking is rennet, which is used to coagulate, or curdle, milk. In traditional cheeses, those deliciously funky ones made in Europe, cheesemakers used rennet that came from an enzyme in the stomach lining of young calves. But, like getting insulin from pigs or cows, getting rennet from calves is far from optimal and, at the time, it was getting expensive. Also, vegetarians were becoming a bigger market group and wanted options that they could eat.

As the industry looked for ways to create rennet in the lab, it settled on insulin as the blueprint for a way to create matching proteins that humans needed, or better yet, desired. What we *want* versus what we *need* is an all-important issue in food technology. We need the basic building blocks of nutrition—proteins, amino acids, and so on. But what we want is often wrapped up in a crinkly plastic bag, and it's full of empty calories that do little for our bottom line.

Pfizer, based in New York City, researched making rennet using recombinant DNA technology. After twenty-eight months of review, in 1990, the FDA decided that this bioengineered enzyme, called rennin, presented no safety risk to the public and could be used in dairy products. That monumental decision opened the door to other foods—tomatoes that ripen faster, apples that don't turn brown—all the way to today, when food-tech startups are actively creating milk proteins, egg white proteins, and, yes, the Impossible burger.

The Chicken-Free Egg

Like milk proteins, egg proteins are on their way to being replicated. In November 2014, Arturo Elizondo attended his first food conference. The half-day event was held at Kaiser Permanente's Garfield Innovation Center in San Leandro, California. While not far from Silicon Valley, San Leandro was known more for being one of the last few industrial towns in the Bay Area. The event itself was tiny, only fifty or sixty people. Josh Tetrick, the founder of Just (at the time known as Hampton Creek), invited Elizondo to the conference, but at the last minute, more pressing matters kept Tetrick from attending. That same week, Hampton Creek was sued by Unilever for using the word "mayo" in its Just Mayo product.* Elizondo looked around for somewhere to sit. For the most part the attendees were from government agencies, nonprofits, and industry locals. "I found an empty chair at the only table with young people at it," he said. "Food tech wasn't sexy at the time."

At the "kids' table" with Elizondo were Isha Datar, the executive director of New Harvest, a research institute dedicated to promoting cellular agriculture; Perfect Day founders Pandya and Gandhi; and Rob Spiro, the co-founder of farm food delivery service Good Eggs. "We talked about tech in food, and using synthetics and what that looked like," Elizondo said. "I told them why people couldn't get all of their food from farmers' markets, which were impractical and expensive—as if the only shoppers were those living in San Francisco." For Elizondo (a vegan), animal agriculture was the ultimate "resource guzzler of the food industry."

Isha Datar is a bit of a celebrity in the syn-bio world. Many point to a paper she wrote in 2010, titled "Possibilities for an In Vitro Meat Production System," as a turning point in growing meat in labs. Elizondo knew of Datar because he had referenced her work in a policy paper he wrote on why China should invest in cell-based meat. Although she eats animal meat, Datar has dedicated herself to promoting cellular agriculture as a way to create a better food system. In 2013, she joined New Harvest as its executive director and, in 2016, New Harvest kicked off an annual conference in Cambridge, Massachusetts. More than a few companies owe their existence to that conference, and to personal introductions that Datar has made. I asked her if she considered herself a matchmaker. "Maybe," she said. "I feel like I'm a natural connector." Listed as a co-founder of both Clara Foods and Perfect Day, Datar was responsible for several other startups, including Memphis Meats and Artemys Foods, coming to fruition.

* Unilever presented that mayonnaise must have eggs in the formulation, and that Just Mayo was effectively lying to the public by labeling it as "mayo" when it did not have eggs. Not only did Unilever eventually lose the case, but it also received a very public outcry against its strong-arm tactics.

Sitting next to Datar at the conference was cell biologist David Anchel, who shared his then outlandish-seeming idea of making eggs without chickens. After they finished talking, the group carpooled to dinner. A few short weeks later, Elizondo, Anchel, and Datar formed Clara Foods. The goal was to use synthetic biology to create the same proteins found in egg whites. For the twenty-somethings, eggs, considered the preeminent clean and functional protein, were the ultimate ingredient to disrupt. And, at the time, no one was working on eggs.

An egg is a mono ingredient. Unlike ketchup, which has several ingredients in it, and is added to many barbecue sauces, for example, an egg requires no helping hands—no binders, gums, or gelling agents. My egg expert here is Sarah Masoni. Masoni is the director of product and process at Oregon State University's Food Innovation Center in Portland, Oregon, and has helped bring more products to market in the United States than anyone else in her field. She understands consumer packaged foods intimately. We met as judges for the annual Fancy Food Show. "Foods taste empty when you try to replace the egg whites," she said. "It can be done, but it will never ever be the same as a product made with an egg." Egg whites are an irreplaceable protein source that are low in fat, high in protein, and no cholesterol.

With that nugget of an idea, Clara drafted a PowerPoint and applied to IndieBio in 2015. For a hot second they called themselves the New Harvest Egg Project. After they were accepted, they dug into how science could answer the age-old question: Which came first? "Instead of raising an entire chicken and waiting for it to lay eggs, only to get the egg white, why not just get the egg white without going through this intensive process?" Elizondo posed.

Unlike most tech founders, Elizondo's expertise was in policy. He worked briefly at the USDA, graduated with a degree in government from Harvard, and interned for US Supreme Court justice Sonia Sotomayor (who has type 1 diabetes). Unlike most Silicon Valley founders, Elizondo, who's from Texas, is typically buttoned up in suits. With smooth-talking investor patter, he rattled off the reasons to make egg protein in the lab, versus on a farm: "They're salmonella-free and hypoallergenic. They have a lower carbon footprint, are sustainable, ethical and, because egg whites are so expensive today, they are incredibly valuable."

The Irreplaceable Egg

He's right. Globally we eat more than one trillion eggs per year—a number that is projected to grow by 50 percent over the next twenty years. Today, the egg industry is valued at $8 billion in the United States. For the month of April, US egg production totaled more than 9.13 billion. When states issued shelter-in-place orders due to COVID-19, wholesale egg prices more than tripled and consumer demand for eggs increased over four times normal levels. This intense market fluctuation is what founders like to point to when they say their methods are "on demand." Rather than wait for a chicken to lay an egg, they can merely ramp up manufacturing.

The egg, while still cheap, is one of the culinary wonders of the world. The proteins found inside an egg enable bakers and chefs to foam, whip, bind, and gel. Despite its flexibility, baking with eggs is an exacting science. If you mess one thing up, you botch the entire recipe. But even when prices double, manufacturers still use eggs. They're irreplaceable.

Or are they?

Today any "bad" actor (chicken eggs, cow's milk) can be replicated in the lab, and when compared with industrial-scale agriculture, these synthetics are deemed by some to be better for the environment. This creation of egg proteins—and dairy proteins that Perfect Day is making—is near identical to the production of rennet and insulin. In shorthand: Genetic code is placed into a "host," which is scientifically engineered using a combination of the right nutrients and living conditions to output the target protein. I'm simplifying this incredibly. (Impossible Foods manufactures heme for its burgers this way.)

But hold on a minute. Do we know the full ramifications of eating these new iterations? We don't. Is there a pause button we can hit so that we can evaluate this new direction? Where's the equivalent of the twenty-eight-month-long review that rennet went through before being approved by the FDA? Are we saying that to benefit the world we need to ditch nature, a finely tuned ecological system that has developed since the dawn of time? It's not this or that—nature or synthetic—but before we embrace another form of industrialized food, let's be confident of the results.

An egg from a chicken contains around forty proteins. In the lab, first Clara worked to determine the ones that would be the most valuable to target for their functional uses in food manufacturing—moisture, thickener, and structure. Next, they had to find the right yeast to become its protein factory. Ranjan Patnaik, Clara's vice president of technology, who joined the company in 2019 after decades at DuPont, walked me through how they build their protein factory. After they "brew" the yeast, feeding it nutrients, mostly sugar and water, proteins are secreted out of the bacteria—this is the end product, but there are still more steps. The yeast, or microbe, will settle to the bottom, and the brew (a clear liquid broth) will rise to the top. A progression of filters catch the protein, and the concentrated proteins are spray dried into a powdered form. If Clara ever achieves large-scale manufacturing, this powdered protein is what Sara Lee might use in its famous pound cake recipe. This is how it will eventually ship to food manufacturers. There are assuredly more steps Patnaik did not include, but his straightforward answer was rare.

Imagine if that Sara Lee pound cake was made without a single egg from a chicken. The basic recipe calls for half a dozen eggs per loaf, but this delicious, moist, dense cake has yet to be re-created for vegans. Our reliance on the chicken both for animal protein and the eggs it lays is immense. Beyond providing decadence to vegans, these synthetic egg proteins could be used to support disaster response teams, and could be widely used in areas that lack refrigeration.

Four years after my first interview with the original founders—I sat down with Elizondo at the Future Food-Tech conference in downtown San Francisco. It was March 22, 2019. We met in a room set up for investors and founders. Neither investor nor founder, I snuck in with my lunch—a big plate of salad greens grown in soil and roasted veggies produced under our ancient sun—and grabbed a table. Elizondo sat across from me, pulling a small vial from his briefcase. This unassuming clear liquid was the tangible result of four years' work, and a reveal I was becoming accustomed to each time I met with founders pushing the boundaries of how food is made.

"This is the equivalent of twenty grams of protein," he said, handing me what quite possibly was worth millions: liquid egg protein, minus the hen. The name for it was CP280. "What does it taste like?" I asked. "Like nothing. You could put it on your salad, in your tea or coffee or in a soda." He didn't offer any to try, but sipping an egg white (on purpose!) was hard to swallow. I briefly flashed to drinking a perfect Pisco Sour, which calls for egg whites. I wondered if Elizondo could make a protein that foamed, but lacked that slightly eggy taste that always turned me off?

Elizondo confidently proclaimed that his team had finally done it: "We're working to launch the most soluble protein in the world," he said. Just a short five years after launching his company, he now had a partnership deal with Ingredion, a multibillion-dollar ingredient producer, and a staff of more than forty employees. Clara Foods was seemingly on its way. All they needed to do was scale considerably beyond the single vial I held in my hands.

A year later, as the pandemic sent us into lockdown, Elizondo kept the momentum going. "We've had the best six months that we've had in the last five and a half years," he told me. Big Food was looking for ways to stay relevant with younger demographics, not to mention "they need companies like ours to be more sustainable," he said. Elizondo shared a long list of announcements that were coming soon, including a name change: Minus Foods. Clever, I said. And hard to resist the potential taglines: "The egg minus the chicken." "All the taste minus the cholesterol." And the ambition of so many New Food founders: "We can work on a minus-egg McMuffin."

Making Whey and Casein from Scratch

Achieving cost parity with everyday grocery store items is imperative for New Foods, but so is regulatory approval. Because Perfect Day was creating something so unconventional—not the process, but the product—the pair began talking to the agency early in the company's formation. "The FDA didn't really know this was possible when the laws were written," Pandya told me. By this he meant that the rules make it hard to use "milk" in the product name, and the path to regulatory approval is unclear.

"This is the kind of idea that goes nowhere in a big company," said Pandya. There are many reasons for that. A robust animal agriculture system exists, and a cold chain—trucks that could deliver milk and keep it cold—to match it. The food industry had spawned proactive lobbying groups that, to protect its product, paid for studies that claimed cow's milk was necessary for children's development, and funded ad campaigns. Remember "Got Milk?" While there are countless non-dairy options on the market, in US schools the only plant-based protein that can be served to kids is soy because it's the closest match nutritionally—it has all the nine amino acids the human body cannot make. This "lack of vision to completely change something as core to food as milk" was a blind spot, said Perfect Day's Pandya. When food regulations were written, the FDA certainly wasn't envisioning this kind of future; and Big Food will certainly try to squash it, invest, or buy it. Pandya hoped his company was nimble enough to stay ahead of them.

The industry's lack of vision is what brings us to where we are today. US sales of plant-based milks rose by 61 percent between 2012 and 2017 according to market research firm Mintel. In 2019, the non-dairy milk sector was valued at $1.9 billion, according to the Good Food Institute. Commercial dairies are consolidating, they're filing for bankruptcy, they're tweeting angrily at journalists, myself included. To compete with the cow, the Perfect Day founders raised millions—more than $361 million by July 2020—and hired almost one hundred employees.

When I last saw Pandya and Gandhi, I dropped by their offices, a two-story Art Deco building in Emeryville, California. Memories of sampling their milk in that Times Square hotel felt like a lifetime ago, not three short years. After talking about funding news and ingredient progress, they broke the news. They finally had a prototype. "Did I want to taste it?" they asked—"it" being ice cream. I thought: I'd eat ice cream made of almost anything, even yeast they scraped off the floor. Sure, I said.

We pushed our chairs out from the sleek glass conference table and walked into a large production kitchen. Jazz was playing from hidden speakers, and a tall man in chef's whites standing at a gleaming Carrara marble countertop. The setup was more fancy cooking school than biotech startup. Pandya slid over two tulip-shaped glass jars. Inside were lab-made proteins. One held whey and the other held casein. In ice cream, whey is indispensable in creating a creamy, smooth mouthfeel. The chef began passing me bowls of ice cream in three flavors: blackberry toffee, milk chocolate, and vanilla salted fudge.

Ignoring the actual flavors, I focused on mouthfeel, texture, and surface appearance. While I ate, I read over the ingredients. Non-animal whey protein was the sixth ingredient after water, sugar, coconut oil, sunflower oil, and Dutch cocoa. It had a slightly different mouthfeel—more frozen yogurt than full-fat ice cream. What the founders told me was that the small amount of whey that was used in the formulation improved everything I was sensing: the mouthfeel, texture, and emulsification of the fat. It was good, but did animal-based whey need an analogue? "It's great," I told them, smiling. "Delicious."

Milk proteins are the building blocks of many of the foods we love: cheese, yogurt, and ice cream. While there are multiple delicious plant-based options, especially the ones with a coconut base, they don't always hit the mark—almond milk is too watery, oat milk is too sweet, pea milk is too vegetal. Perfect Day spent almost five years looking for the right microflora (or yeast fungus—a phrase that neither founder relishes) that they could engineer to brew milk proteins. When they found it, they nicknamed it "Buttercup."

Once Perfect Day's yeast has been encoded, it's placed inside fermentation tanks. At this point, the conditions must be ideal in order to fuel growth and generate proteins. To get a handle on the process, I interviewed Tim Geistlinger, Perfect Day's chief technical officer. "Proteins are made of carbon, nitrogen, and oxygen," he told me. "The basic things that people look at [in fermentation] are temperature, oxygen, and stirring rates. That allows for the oxygen, it allows for transfer rates and how fast we feed them sugar and nitrogen." I nodded along to his explanation. It synced up to so many of the other foods in my book—key ingredients (often proteins) taken from nature and replicated in the lab. Was this further industrialization of our food system a good thing? Instead of pushing for greater biodiversity, farmers would continue planting wheat, corn, and soy in order to support our new non-animal wheels of cheese.

In the early days when Clara and Perfect Day attended cellular-meat happy hours, they spent their time talking to each other instead of networking with everyone else. Somewhere there was a line that was drawn. Elizondo said he thought it was because what they were doing wasn't "as sexy as meat." And he may be right. Where meat has dozens of startups vying to be first to market, eggs and dairy still only have a handful trying to reverse engineer our breakfast staples.

Geistlinger has worked in food before. Previously, he was at Beyond Meat, helping Ethan Brown create his now best-selling plant-based burger. Prior to Beyond, he developed malaria drugs at Amyris—Neil Renninger's company, which shows up on a surprising number of food-tech résumés. Where scientists once had straightforward career paths—typically either academia or pharmaceuticals—now there is a third possibility in food tech, which gives them an opportunity to make millions, while, as Geistlinger suggested, "impacting the world in a positive way."

Making foods at bench scale—a term used to define the small amounts produced in a lab—isn't necessarily easy, but it's far easier to make enough tiny vials to stash in a backpack than it is to fill a 200,000-liter bioreactor or a food-grade tanker truck, let alone enough freight containers to fill the hull of a ship. One hurdle after dealing with regulatory approvals from the FDA is sourcing cheap sugar, which is the most crucial factor in growing any of these protein analogues. In my research, I inquired about the sugar source, and whether the origin matters—food waste, corn sugar, sugar beets—and whether the sugar, or the quality of the feed source, is something that is carried over to the new protein. Their answer was that the sugar source doesn't matter. That it's completely "used" up by the bacteria in the formation of these novel proteins. An NYU chemist that I've interviewed in the past liked to say to me: "Garbage in, garbage out." I want to believe the founders, but because my life revolves around thinking about sugar, which in the body is a carbohydrate, I approach these advances with a certain amount of trepidation.

Scaling Beyond the Lab

Once Perfect Day had successfully made small amounts of the ingredients in-house, it needed to simplify the steps in order to scale it. For comparison, it's like attempting to cook the most complicated recipe you've ever seen, perhaps from Ferran Adrià's cookbook, *El Bulli*, named after his famed molecular gastronomy restaurant. Then, once you make the recipe for a dinner party for four, you have to make it again for one hundred people, then one thousand, one hundred thousand, and finally one million. To hit Perfect Day's goal of ending our reliance on animal agriculture, that one million is merely a starting point, because the founders want to open factories all over the world producing casein and whey for billions—both people and dollars.

Similar to Clara's path with Ingredion, in order to commercialize its animal-free dairy proteins, Perfect Day signed an agreement with ADM, which has sales of more than $64 billion annually. This partnership with Big Food allows Perfect Day to one day (potentially) scale its dairy ingredients. For ADM, the company gains access to much-needed innovation. It also ties Perfect Day's fate to a multinational company that's been accused of price fixing, and spends money supporting industry spin about food safety and nutrition, which includes promoting and defending sugar, artificial sweeteners, food additives, and pesticides.

In 2018, Perfect Day was awarded its first patent: patent number 9,924,728 for "food compositions comprising one or both of recombinant beta-lactoglobulin protein and recombinant alpha-lactalbumin protein"—essentially, whey and casein.

In June 2019, Perfect Day filed an application with the FDA for something called GRAS, which stands for "Generally Recognized As Safe." According to the FDA's website, "any substance that is intentionally added to food is a food additive that is subject to premarket review and approval by FDA, *unless the substance is generally recognized, among qualified experts, as having been adequately shown to be safe under the conditions of its intended use*." (Italics are mine.) What this means is that only the unusual ingredients are reviewed; anything common (such as sunflower oil or Dutch cocoa) is given an automatic pass into our food system. When a company applies for GRAS, it shares its own research versus that of unbiased parties. The problem lies in the phrase "qualified experts," which typically means scientists who are paid for by the same company seeking GRAS. This is the regulatory path that Impossible Foods used when it filed an application to have its heme—the sine qua non ingredient of the Impossible burger—labeled as GRAS.

Because rennet and insulin have been made using processes like these, and milk has been consumed for centuries, these non-animal-derived milk proteins can slide in under the regulatory wire with little to no oversight. Ditto egg proteins. This simplistic way of thinking is misleading, and it's unfortunate that GRAS has become the de facto regulatory route for many of today's food tech companies; and also, unfortunate that the FDA no longer applies the same rigor in its review of novel additives getting adopted into our food system. This doesn't mean that I think they're doing something risky or purposefully unsafe, but I would like to see a deeper review that includes academic studies done by independent food scientists without any financial interest in the final product. As of this reporting, almost no food tech companies are doing studies like this.

The acceptance of milk spouting Willy Wonka–esque from inside tanks, rather than cows, may take time. A 2017 survey by the magazine *Vegetarian Times* found that 7.3 million Americans follow a vegetarian-based diet. One million of them are vegans. There are 327 million people in the United States, which means only roughly 3 percent of the population is eating a solely plant-based diet. While much of the data points to industrial agriculture as being far more detrimental to the environment, the trade-off here is that we're moving to another version of industrial manufacturing—just one that no longer includes animals. In this new version, we still need factories, energy, water, and crops. These solutions seem functionally safe, but we won't truly know for decades to come.

"Manufacturers are a little skeptical of this Frankenstein stuff," said Masoni, the OSU food innovation expert. As long as the egg industry stays stable there may not be a big enough push to make this change, and as Masoni pointed out, it's a huge lift for someone to change an ingredient. "I don't know how long this will take to be implemented and accepted, maybe twenty years?" she wondered aloud. When we finally get to execution, scaling to the point of being in every supermarket, "everyone will be thinking about what's next," she said.

The Fastest Way to Consumer Adoption: Ice Cream

"Non-animal whey protein" is a bit wordy for an ingredient label. But Pandya of Perfect Day thought it was the most transparent to consumers. "Plant-based is more confusing," he said. Even without the cow, Perfect Day's ice cream needed an allergy warning because the proteins are near identical. The only difference is that Perfect Day's whey doesn't have the lactose* found in cow's milk. Several studies over the last few years have reported that more than half of consumers review the ingredient label

* Lactose is the sugar present in cows' milk and dairy products, and it's often what people are allergic or sensitive to.

as the most important educational element when they are shopping. I wondered if these four words would raise any red flags?

When I posted a picture of Perfect Day's ice cream to a vegan business organization I belong to on Facebook, the more than nine thousand members-only group was confused and mostly put off by the concept. "It contains milk proteins," one person commented. "Yes," I replied, "but not from a cow." That didn't help.

One woman posted: "They have to explain that 'contains milk protein' on the label doesn't mean cow milk. I'd put it back on the shelf otherwise."

Another wrote: "Interesting! I'm not sure I would buy it as a vegan. I would want to do more research. I think it could be marketed to vegetarians and meat eaters who want alternatives. I think the label really should make a person picking it up understand it's a new technology, and why it's different from the rest."

Yet another commented: "Since no cows are involved, I'm fine with it in terms of ethics/veganism. But it doesn't pass my health standards. I'm hoping someday we can have lab-grown dairy without cholesterol or saturated fat. There is no cholesterol, but a lot of saturated fat in this, because the third ingredient (after water and sugar) is coconut oil."

These comments came from a highly engaged group with a specific agenda—staunch vegans—and shows only that there's an uphill slog toward ensuring that the marketing of these products contained the right amount of education and transparency. However, niche groups (sorry, vegans) mean little to these food tech founders. For them, it's all about mainstream adoption.

When I decided to write this book, one of the areas I wanted to understand better was the postprocessing of these high-tech fermented proteins. These could include the original host yeast (that was GMO), the protein, plus various other impurities before being purified with an industrial solvent, which no manufacturer has been willing to identify. Was it possible that the end ingredient contained extracellular bits? Is it possible down the road that the FDA will be lax in its oversight, for example, when the ingredient is being made around the clock in factories around the world? Will regulatory officials inspect the ingredient before it's formulated into something else, such as baby food?

Younger consumers today seem to be clear that the cow-based supply chain comes with big compromises—there are safety issues, health concerns, and environmental factors. But it's the devil we know, right? In our future supermarket aisle where the options include cow milk, non-animal cow milk, and plant-based non-dairy milk, how will we decide? If the plant-based product is decent, why opt for the high-tech version of cow's milk?

Because I'm not a trained scientist, I took my questions to Cesar Vega Morales, a food scientist with a PhD from University College Cork in Ireland. When I asked him what type of education was needed for consumers to better understand an ice cream containing milk proteins that did not come from a cow, he said, "Personally, I think it's inconsequential: the molecule is the molecule. Why do we need to talk about it? It's dairy-free and that's the end of it. Sometimes we get too hooked up in the technicalities. Sometimes consumers don't need to understand it."

People today may want to know the page-turning story behind futuristic foods, but few will be driven to go deeper. I'm different. I want to know what's in my food at a granular level. A meal based on whole foods versus its processed counterpart (and with identical amounts of carbohydrate, protein, fats, and fiber) will necessitate completely different amounts of insulin.

Later, when I asked Clara Foods' Elizondo whether an ingredient company owed it to today's better-informed consumer to be more transparent, he assured me they "cannot ignore consumers" and that he was "a firm believer that B-to-B companies of tomorrow can't look like B-to-B companies of yesterday."

Neither Perfect Day nor Clara Foods has scaled their production yet, but both companies are now backed by massive ingredient companies that will surely take them there. When I spoke with Victoria De La Huerga, vice president of ADM Ventures, a major investor in Perfect Day, she told me it wouldn't be easy to make the quantities Big Food demanded. "With scaling it's always how can you do it as inexpensively as possible. That takes a fair amount of engineering help," she said.

When they do get there, let's hope that transparency is not lost in the interest of moral imperatives and the need to make money. Food scientist Morales was more cynical: "Consumers aren't looking under the hood. If they were, they would see all of these ingredients* that are processed in so many ways. If they knew a little more they would be surprised. This is the nature of the consumer. Who cares?"

I laughed along with him, and at the same time I cringed. In another vote of confidence, Bob Iger, the former CEO of Disney, had recently joined Perfect Day's board. I'd interviewed Iger in the past, and I knew that the businessman had a soft spot for ice cream. In addition to Iger, Perfect Day's "non-animal whey protein" had begun appearing in other commercial brands, including Brave Robot and Nick's Ice Cream.

Back in Perfect Day's soothingly perfect kitchen, jazz music lulling me into acceptance, I licked the spoon slowly and let the ice cream melt on my tongue. It didn't taste as rich as Häagen-Dazs, but it also didn't taste icy like many non-dairy brands. It was creamy and stayed long enough on my palate to give my taste buds the message: ice cream—yay! It was delicious. If I had tried it in a blind tasting with other ice cream, I would have no idea that it wasn't traditional dairy. A few weeks after I tried the three scoops, Perfect Day sold its first one thousand pints for $20 each, or $60 for all three flavors. It sold out in less than half a day. Sure it was good, but it was expensive, and sugar was the second ingredient. It still spiked my blood sugar.

* Everyday ingredients can be found in your favorite breakfast cereal and include maltodextrin, dextrose, and soy lecithin, for example.

Chapter 5

Upcycling

Rescuing Edible Stuff to Make . . . More Edible Stuff?

Lost Food

When I lived in Manhattan, I stored my compost inside my freezer. Jammed into old plastic bags you'd find frozen blocks of carrot peels, avocado pits, apple cores, and coffee grounds. I'd put off my trip to the Union Square farmers' market for as long as possible. When I ran out of room in my freezer, I'd weigh down my bicycle with massive totes, one bag in my front basket and the other two hanging from my handlebars. On my wobbly ride to Union Square, I'd pray for a stream of green lights so I wouldn't have to stop my overloaded bike. Finally, after dumping the contents into gray trash cans manned by the Lower East Side Ecology Center, I'd allow the warm feeling of being an environmental do-gooder to swell.

Then I read journalist Amanda Little's book *The Fate of Food* (2019). In it I learned that my sense of virtuousness was misguided. Little interviews Darby Hoover, a waste researcher with the Natural Resources Defense Council (NRDC), an environmental advocacy group. According to

Hoover, the entire category of food waste is "riddled with contradictions." Contradiction number one was that healthier diets are often the most wasteful—as evidenced by my constantly filled freezer. "Not generating waste is far better for the planet than recycling food scraps." There went my Girl Scout sustainability badge!

My interest in upcycled food probably came about because my other passion (compost) is completely uncool. Food waste is icky. It brings up feelings of guilt, reminding us of food that we spent good money on and failed to use. Upcycling* in the food industry—the art of capturing still-nutritious waste from one process to create entirely new edible products—allows us to feel virtuous. I eagerly followed the category along, writing stories for the *Wall Street Journal* and *New York Times* about the people leading the efforts to capture waste streams.

One of the first waste products to be magically turned into something else was oleomargarine, a name that was eventually shortened to just margarine. Originally crafted out of beef fat for Emperor Napoleon III, it later became a way for US meatpacking plants to deal with their excess waste. By the early nineteenth century there were dozens of companies making this butter alternative. Before getting the do-good name of "upcycled," these foods were referred to as co-products or by-products. Whey, a leftover liquid from yogurt and cheesemaking, is a much more successful example of rescuing a waste. Tapped first in the eighties, whey was easy to digest and high in protein—it made sense as an animal feed, but it's worth much more for humans. Today, it's formulated into hundreds of meal-replacement bars, protein shakes, and keto crackers.

"Back then we didn't want to call it waste because people wouldn't want to eat it," said Tara McHugh, center director at the USDA's Western Regional Research Center, one of the department's four national centers (formerly called "utilization labs"). Today, people don't mind eating waste. In fact, it's becoming something of a moral imperative for manufacturers

* An example of upcycling in industrial processes is asphalt, which is a biproduct of the refining of petroleum, or crude oil.

to convert surplus foods into value-added products. While some waste streams can go to animals, keeping food as food for humans is higher up on the EPA's food recovery hierarchy—a ranking system that shows what is the most beneficial for the environment, society, and the economy.

At the USDA, McHugh works with companies to create new foods from agricultural waste, such as pomegranate peels and the pomace from winemaking. With large companies announcing their sustainability commitments, and more marketing dollars being spent to educate consumers, McHugh said that "it may be more clear to consumers" that what we're buying is beneficial for the environment. In a 2017 Drexel University study on the benefits of upcycling, researchers noted that "consumers forgo personal gains if they feel that purchasing pro-environmental products will contribute to the welfare of society."

Other research backs this up. At the annual Institute of Food Technologists meeting in Chicago in 2019, Mattson, a Bay Area company that helps clients develop new foods and beverages, released data showing that 39 percent of consumers aim to buy more food and beverages using upcycled ingredients. That number is expected to rise to 57 percent in 2020. ReFED, a Bay Area nonprofit that uses data to lobby and organize companies around food waste, counts at least seventy US companies transforming wasted food into new products. The Upcycled Food Association—yes, there is an actual organization—supports industry growth along with setting guidelines for what can be considered upcycled, and counts north of ninety companies globally as members.

Even during widespread COVID-19 shelter-in-place orders, consumers ranked sustainability high. According to a study of two thousand adults by Genomatica,* 86 percent of respondents polled said that they would continue to consider the importance of sustainability even after the pandemic subsides. Additionally, 37 percent of Americans are willing

* The sample used by Genomatica was balanced by census targets for age, gender, and geographic area and has a margin of error of about +/–2 percentage points. Data was collected from June 16 to June 24, 2020.

to pay a little more for sustainable products, even during an economic downturn. Across age groups, Gen Z is the most willing, at 43 percent.

While we appear to be getting more virtuous in our shopping habits, it's worth taking a moment to roll back the film. Once again we can look to a post–Second World War America and our advances in chemical inputs as the watershed moment. When factories stopped producing wartime ammunition they opened the door to a new industrial food system spurred on by synthetic fertilizers and large-scale production. After the lean wartime years, and the growing middle class, came a matching reflection of plenty on our grocery shelves. The list of "advancements" from this time is long. A few: frozen convenience foods, plastic-based packaging, easy access to refrigeration, economies of scale, and governmental subsidies to large-scale agriculture. All of these fostered the growth of industrial food, which then resulted in industrial-scale waste.

Reducing food waste isn't as simple as tossing a few carrot peels into a chicken stock. As a culture we've spent decades buying perfectly round, blemish-free apples and carrots without wonky bumps. Outside of a few pandemic surprises—runs on toilet paper and hand sanitizer—we expect to see our supermarket shelves laden with goods. But peek in the trash bins behind the store and you'll find all of this edible, nutritious food in the trash because it's less than perfect: dairy products with a best-by date past its prime, day-old baked goods, brown bananas, and overripe berries.

After decades of being ignored, food waste is officially top of mind. Innovative chefs created pop-up restaurants themed around "wasted food," one-off ticketed events were organized around showcasing food supply inefficiencies, and for-profit companies were created by do-gooders eager to utilize hidden elements of our food system. In its trend report for 2021, Whole Foods noted "a huge rise in packaged products that use neglected and underused parts of an ingredient as a path to reducing food waste."

Terminology varies for this tossed treasure. "Wasted food" is what the NRDC calls it in an effort to "signal a shift in thinking by indicating that it's good food, not trash." I prefer the term "lost food," used at Expo West by Walter Robb, former co-CEO of Whole Foods. An upside of this term is that it shifts the implied blame in the word "waste" to the more empathetic word "lost." Lost means we can find it again. It can be rescued and reinvented.

Industrial symbiosis, the technical name for this field, is indeed good for the planet, but it doesn't address our tendency toward overconsumption, and food manufacturers' predilection to support and urge it on. It doesn't lessen the inequality in our pantries: those of us with a year's supply of snacks in our cupboards and others with barely enough to get through the day or week. Whether products made from upcycled ingredients are healthy, or whether they perpetuate an American diet reliant on snack foods—and our hardwired brain chemistry that can't resist crunchy, salty, and fatty—remains to be seen. With near matching concerns, and a fun person to follow on Twitter, *Washington Post* columnist Tamar Haspel told me she, too, was "skeptical that we're going to get healthy foods out of those waste streams."

Eat Beer

The first step to making beer is to add a heap of grains, often malted barley, into a tank along with water. This mixture, called the mash, is heated, helping to break down the cell walls of the grain and release sugars. This will eventually convert into alcohol. After a few hours, the grains—called BSG for brewers' spent grain—are dumped. When possible, breweries send this wet mass off to farms as a livestock feed supplement; more often, though, it's thrown away.

The idea to recycle spent grains from breweries and distilleries originated in 1913. Jean Effront, a Belgian chemist, suggested that new foods with a strong "meaty flavor" could be created from brewers' and distillers' waste that were three times as nourishing as beef. With impressive foresight, Effront even noted that "near meats made from this substance would be more economical because they would not need to go through the intermediary of the animal that transforms them into meat." This is the same argument used by today's advocates of plant-based and cell-based meat.

Across the United States, several companies are searching for a use for these grains beyond animal feed, including Rise, Brewer's Crackers, Grain4Grain, and NETZRO. These entrepreneurs have tried everything: They tossed it into bread, turned it into baked cereal, and made cookies. But there were problems. Some products were not edible. Worse, the ones that were edible were yucky. The problem was that the semi-hollow grains often showed up where you didn't want them, like between your teeth. To create a palatable end product, the grains need a further level of processing. This involves drying BSG at high temperatures—some companies use infrared technology to kill pathogens—and milling it into a fine powder.

Dan Kurzrock, whose company motto is: "Eat Beer," calls the powder he makes Supergrain+. Kurzrock is the CEO and co-founder of ReGrained, a startup based in Berkeley, California. I had an affinity for Kurzrock because he biked everywhere. In another life he would make a great park ranger. Like many good ideas (and some bad), the concept for ReGrained began over beers. Kurzrock and his co-founder, Jordan Schwartz (who is no longer with the company), met in Hebrew school and learned to brew while attending the University of California, Los Angeles. After making a six-pack, the pair had a pound of what looked to them to be still-nutritious food. Their minds jumped from that tiny amount to the millions of tons of waste coming out of big-name brewers like Budweiser, Miller, and Molson Coors. The latter is now an investor in ReGrained.

No one wants to acknowledge how big the waste stream is. While the brewing industry has member associations in spades—including one for master brewers, one for brewing chemists, regional associations, and a national association—no one is tracking how much BSG is tossed or diverted to farms. I did a back-of-the-envelope calculation using statistics from the Alcohol and Tobacco Trade and Tax Bureau (TTB). In 2019, there were more than eight thousand breweries producing almost 191 million barrels of beer. As a benchmark, the US Brewers Association estimates that roughly seventy-two pounds of malt are used per barrel, but this is a dry weight. Wet grain is heavier, but much of that is water. Another factor here is that small craft brewers typically use three to four times more malt than large beer makers. If we ignore that and just use the seventy-two number, we get almost fourteen billion pounds of spent grain for one single year. A portion of this spent grain goes to farms, but no one is tracking how much, and the rest goes into the waste stream. Depending on the location of the brewery, it might be composted.

The distinctive musty brewery smell is not unlike the yeasty smell of oatmeal bubbling away on the stove, minus the sweetness. When dried, the spent grains look like fragments of cooked brown rice. Most of the sugar is gone, which makes the flour different from our usual white flour. But Kurzrock was right, it is still nutritious. A cup of Regrained's BSG-based flour has 3.4 times more fiber than whole wheat flour, as much protein as almond flour, and includes iron, manganese, and magnesium. Nutrition stats like these are compelling, but ReGrained's flour doesn't work well as a flour in its own right. It's best when formulated into other foods, which leads us to snack town.

For snack makers in 2020, the puff appeared to be the shape of choice. There are now puffs made of chickpeas, cleverly called Hippeas; there's an array made from green peas, including PeaTos, Peas Please, and Harvest Snaps; and there are paleo versions made from cassava, a starchy root vegetable. ReGrained makes granola bars, and now they have puffed snacks made from corn and its BSG flour. Kurzrock told me they used non-GMO corn "because it tasted better." And they are tasty, but chips, including puffs, are bound to hit my blood sugar quicker than I'd like because they're made using an extruder, the giant machine I mentioned earlier that shoots out shaped, cooked easily digested food.

Nutritional studies about extruded food, which is by definition highly processed, are clear. Increased processing leads to increased glycemic load, which is another way of saying that your body gets hit with a fast glycemic spike. The University of Sydney, Australia, has a group that looks solely at the glycemic index of foods. They note that high glycemic foods "are digested in a flash because the production process makes the starch very accessible." Management of blood sugar levels is crucial to people with diabetes, but doctors will say that ping-pongs in glycemic response are good for no one. The quality of foods we eat also translates to hunger levels later in the day, as pointed out by the clinical nutritionist Dr. Michael Greger on his website NutritionFacts.org. Eating quick oats, he writes, will lead you to be hungry sooner, and to eat more during the day than if you had eaten steel-cut rolled oats, which are far less processed.

An expert in the land of upcycling is Jonathan Deutsch, a professor of culinary arts and science at Drexel University. We met while judging the Specialty Food Association's annual fancy food show. This task had highs and lows: from sampling twenty-five kinds of chocolate bars to dipping my fork into fourteen salad dressings. Deutsch is also the director at Drexel's food lab and works with dozens of companies on product development. Upcycling is his sweet spot. "Here's my position: Most food that we eat is processed, even produce. We need to talk not about unprocessed and processed, we need to talk about ultra-processed." For Deutsch there was "opportunity in all areas of the food system." It all came around to what these makers were doing with their upcycled ingredients, and how much of it was going into the final product.

While ReGrained is making bars and puffs, Sue Marshall, founder of Minneapolis-based NETZRO, is making pancakes. The COVID-19 pandemic supplied the need. "I had a few people reach out to me saying there was no flour," Marshall said. She quickly collaborated with a local miller to make a rye and wheat blend with 20 percent BSG. Another ingredient for Marshall is eggshells, which are almost universally wasted but can be salvaged and turned into calcium and collagen. Marshall sees limitless potential in upcycling—if only she had the money. "Investors want us to focus on one thing," she said. "But entrepreneurs don't want to do just one thing. I'm a woman, I can do everything."

Unlike most other food companies, ReGrained is conducting third-party tests on its ingredient. The founders applied for a small business loan and partnered with the USDA to run a nutrition study. Phase one was an animal-based feeding study, and ReGrained co-founder Schwartz shared over email some of the preliminary results. The "potentially positive results" included possible reductions in cholesterol levels, and increased levels of microbiota in the guts of the animals tested. The human feeding test, which they need additional funding for, will be in a second phase. Based on the animal study, Schwartz hopes to see high levels of prebiotic fiber—usually linked to the vitality of the gut microbiota—in ReGrained's flour.

ReGrained also partnered with the USDA on the technology to dry and mill BSG. They filed a patent for the process and are building a small factory in Berkeley, California. Once up and running, ReGrained can go from making one ton of flour per week to a ton per hour. "We have plenty of supply," said Kurzrock. His list of possible suppliers is almost limitless, but currently he works with Fort Point Beer Company, which makes its beer in a fourteen-thousand-square-foot brewery in the San Francisco Presidio. "They're ideal for us," said Kurzrock. To date the startup has raised more than $4.2 million in funding. Barilla, the Italian pasta company, is an investor, and is actively testing ReGrained flour in dried pasta. Griffith Foods, a hundred-year-old ingredient company, is testing it, too. There are other third parties playing with the flour, but all are subject to nondisclosure agreements (NDAs), and information wasn't forthcoming.

This secretive business model, which is pervasive in the food industry, relies on the obfuscation of ingredients, processing, manufacturing, and distribution. It needs to change to keep up with the times—sharing via social media is everything. Even though younger consumers, millennials, and Gen Z are expecting more transparency, Big Food continues to conceal—its agenda and methods—and exerts pressure on smaller entrepreneurs to follow suit. My book isn't a primer on how we can improve our food system, or only a rallying cry against industrialized food, but an awakening that New Foods are following in Big Foods' path, taking their investment money and (even) being acquired by these legacy brands. (A narrative that occurs in all areas of technology.) To be a little cutesy, my beef isn't with these small companies, it's that they create a virtuous path for Big Food that leads us straight to convenience foods, and the snack aisle—home of cheap, low-quality calories.

The Early Days of Upcycling

In the late 1950s, "spun protein" was the future of food. (I covered Robert Boyer's invention of turning plant components into industrialized food in Chapter 3.) In addition to using virgin materials like soybeans, Boyer figured he could capture waste destined for livestock or the garbage heap. He was particularly interested in the defatted meal of oilseed crops like peanut, safflower, and alfalfa. Even in its early days, General Mills saw the importance of rescuing food waste. Embedded in its versatility and potential, and the reason to invest so heavily in the process, was what historian Nadia Berenstein called the "world-saving discourse," which centered around the priority protein gets in human nutrition and the ways we have become overly invested in it. "It gets more attention than it deserves," she said. "It's all about this performative virtue—it's metabolic fuel, masculinity, strength and muscle building."

In 1965, the head of General Mills' Isolated Protein program projected that wasted protein sources found in underdeveloped countries could be tapped to fill missing nutritional needs. Despite the promising path—or perhaps yet another version of America's ongoing destruction of indigenous food culture—neither upcycled projects nor spun protein made it past the 1960s. However, the beneficiary of these advances in food tech and industrialization is the snack food category, which got its own aisle in the grocery store beginning in the sixties.

If Boyer were alive today, he might be eagerly eyeing the pomace left over from pressing canola and olive oils. But there are still entrepreneurs who believe in soy, despite the ambivalence many Americans have toward the ingredient. Claire Schlemme has zero ambivalence toward soy. We met in 2016 at the Good Food Spotlight in Manhattan, a monthly networking event created by food entrepreneur Rachna Govani. The spotlight evoked *Shark Tank,* minus the big money and snarky attitudes. I attended as often as I could. Sponsored by Foodstand, Govani's then startup, and the occasional big brand, the event gathered a trio of experts to listen to pitches from early stage food companies. After each pitch, judges shared their feedback on topics ranging from pricing to taste and packaging. A crowd of foodies watched the action, voting with their mobile phones, which were tallied on a wall behind the judges.

At the event, Schlemme had a plastic container filled with what she called "wasted" cookies. The base for the cookies was okara, the fibrous pulp left over from tofu making. I was intrigued. Already determined to shrink my own garbage footprint, I enthused over her idea. At the time, Schlemme and her co-founder had a company name, Renewal Mill, but little else. She waved me off: "We're not ready for press."

Tofu has been consumed for millennia, and in Asia, where its culinary use began, okara is not discarded. But in the United States, what didn't go to animal feed was tossed. After taking some of the waste from a local tofu maker, Schlemme and her partner dried and milled it into a flour-like substance. In one cup, the final product has 47 grams more fiber than one cup of white flour, plus more protein and fewer carbohydrates. Because it's made from soybeans, okara has all the key amino acids necessary for humans—the ones we can't make on our own. It's highest for L-glutamic acid, or glutamine, an important amino acid utilized in the body for many functions. During intensive exercise, glutamine levels can drop. In 2015, a study performed on college athletes who ate okara-based products—two cookies a day for six weeks—showed that "markers of fatigue and muscle damage were significantly decreased."

Many upcycled flours are low in gluten, but okara has none. Without gluten's rising properties, don't expect it to make a fluffy biscuit. For a holiday party last year, I made a hachiya persimmon pudding using a blend of okara, almond, and white pastry flours. It was custard-like because of the fruit, and higher in protein than a normal dessert. It didn't look very pretty, but everyone at the party said it was delicious. Little did the partygoers realize that they were also getting more calcium and fiber—the number one missing nutrient in the American diet—thanks to my dessert. Upcycling for the win!

Initially, Renewal Mill was working on the East Coast, but it relocated to the San Francisco Bay Area in 2018 and signed a partnership agreement with Hodo Foods in Oakland. Minh Tsai, the CEO and founder of Hodo Foods, is another firm believer in the magic of soybeans, but above all he's a businessman. "Tofu has this history of a thousand years of efficiency," he said. At Hodo, Tsai is attentive to sustainability. "Are we wasting water? Are we wasting any by-products like the okara? Are we wasting any inventory?" Instead of a linear economy, Tsai believes in a circular economy—a regenerative form of business practice that aims to do away with industrial waste and the depletion of finite resources. In Europe, this type of sustainable growth is part of a clear agenda toward becoming the first climate-neutral continent by 2050.

"Even before I made a block of tofu we knew we would have [okara]," he said. And he knew he "didn't want to throw it away." Since founding Hodo in 2005, Tsai has been selling okara off as an animal feed, which was then the easiest solution. When Renewal Mill came along in 2018, Tsai decided it was time to shift to an ingredient for people, a supply chain with more value to his bottom line. In Oakland, California, Renewal Mill is setting up a pilot production line inside the Hodo factory that picks up where the tofu process ends.

Renewal Mill's office is just down the street from Hodo's bustling tofu factory. I dropped by on a meltingly hot day, and their front door was wide open to the street. Instead of a door, there was a pallet blocking off the entryway because crawling on the floor was a plump, happy baby. This was Arlo, Schlemme's ten-month-old son. While a nanny spent time with Arlo, I pulled up a chair around a homey conference table set with baking tools jutting up from silver pails, and talked about the future with Schlemme and her COO, Caroline Cotto.

We talked about how soy goes in and out of popularity, plus how it's one of the top eight food allergens. For these reasons, Renewal Mill is looking to break away from being known solely for its okara flour. Schlemme and Cotto are investigating other supply chains with compelling waste stream stories like what's left over from making vanilla and the now hugely popular oat milk. As with ReGrained's Supergrain+ products, there are limited okara-based products on the shelves nationally. If you're in the Bay Area, you can find Renewal Mill's dark chocolate brownie mix sold locally. Online you can find the flour and brownie mix, cookies, and gluten-free tortillas from Tia Lupita made from a blend of cassava, cactus, and okara flours.

Squeezing the Juice

It's hard for me to find flaws in these upcycled entrepreneurs. They are universally passionate about doing good by the planet, and they are all exceedingly nice. When I had questions, they answered them as best they could. If the Drexel study found that people will purchase an upcycled food at a higher rate than virgin ingredients, then I was in their crosshairs along with so many others. In 2020, upcycling was estimated to be a $47 billion industry.

Long before that, in 2014, I wrote a story for the *Wall Street Journal* about Salvage Supperclub, a dinner party where guests sat in a dumpster and ate a meal composed entirely of rescued ingredients. Not realizing where the photographer would stand to take his photographs, I sat front and center, and wound up with my picture in the newspaper for the very first time. (Hi, Mom!)

A year later, chef Dan Barber announced he was turning Blue Hill, his Manhattan restaurant, into a pop-up called wastED, featuring only foods that could be salvaged. Barber covered the walls in canvas fabric and invited guest chefs to join him in dreaming up all manner of high-end dishes made with low-end ingredients, including skate wing cartilage, pasta trimmings, and fish collar. All these scraps highlighted how easy it is to waste perfectly good food. My iPhone photos from this night are dark, lit only by homemade candles made from rendered beef fat that we dipped our bread into and wasn't actually gross.

According to a 2019 IBIS World report, revenue from juice bars in the United States was expected to hit $2.7 billion in 2020, which is a

1.9 percent rate of growth.* Estimates of waste coming out of the juicing industry aren't easy to find, but one can assume it's several hundreds of thousands of tons annually.† Unfortunately, when we waste ingredients, we also waste the resources that went into creating these products—water, energy, labor, ingredients, plus any that are needed to grow fruits and vegetables, such as soil, seeds, and fertilizers.

A burger made from juice pulp sounds pretty dismal, but it was the dish that got me most excited at my wastED dinner, and it was made from pulp that Barber's team had rescued from Liquiteria, a New York City chain of juice bars. The dense and savory burger rewarded my taste buds with fattiness (from cheese, almonds, and canola oil) and heartiness (from beans and mushrooms), and it was wholly satisfying because it also had protein (from tofu and eggs). That burger stayed in my mind as a product that could show leaders in the food industry that consumers are looking for foods that are better for the environment *and* healthy and delicious. Shortly after that memorable meal, Shake Shack announced that it would be selling the pulp burger for a limited time. As soon as I heard, I took the subway to the Gramercy Park Shake Shack and waited in line for an hour. When I got to the front they told me the burger had been made for only one day, the day before. Crushed, I walked away empty-handed and hungry.

As a conduit for selling novel foods to Americans, the burger is it, as evidenced by the stampede Big Food is making to compete with the meat-free meat-like burgers from Impossible and Beyond. Despite this, and despite the quick sell-out of the Shake Shack pulp burger, the reuse of pulp still languishes in the bin of good ideas. This means that juice pulp—with its precious fiber—was begging for a solution beyond trucking it out to farms as animal feed.

Waste streams like this are what opened the door for Kaitlin Mogentale. She started her company, Pulp Pantry, after seeing the problem

* This data is from February 2019, and we can assume that due to the coronavirus pandemic the actual revenue numbers will come in lower.

† In 2016, *Modern Farmer* reported that 175,000 tons of juice pulp were sent to landfill.

firsthand in Los Angeles, a city that is notorious for its juicing habit. "I saw so much waste coming out of juicing," she said. The idea was almost handed to her on a compostable bamboo tray. But how to turn it into a business? Initially, she considered creating healthy food for kids. Then she made granola from the leftover pulp and sold it at her local farmers' markets. With that early product, Mogentale applied to incubators and accelerators. She joined Food-X in New York and later she landed a spot with an incubator run by Target. At the end of 2019, Pulp Pantry launched its first commercial product: upcycled tortilla chips.

Corn is still top dog in chips. But if you walk the snack aisle of your local market, you'll notice there's competition. Stacked as high as the eye can see are chips made from trendy new ingredients like cassava, garbanzos, egg whites, and chicken. Yes, chicken.. The "salty snacks category" is growing. In 2019, it was up 4.9 percent to $24.9 billion, as reported by IRI, a CPG analytics firm. Tortilla chips were also up 4.9 percent and were valued at $5.5 billion.

But instead of corn, Mogentale's chips are made from a mix of kale and celery fiber pulp directly sourced from Suja, a $100-million-plus juice company located in Oceanside, California. According to Mike Box, the COO of Suja, it sends approximately seven million pounds of pulp waste annually to nearby farms as animal feed. But according to the EPA food recovery hierarchy, feeding hungry people is higher up on the pyramid than feeding animals. (Source reduction is at the very top.) Suja now delivers a small fraction of its waste to Mogentale as frozen pulp. This helps retain its freshness until it's dried and milled into flour. The chip is a blend of Renewal Mill's okara flour mixed with chia seeds, cassava, and tapioca flour. "I love the focus on fiber. I think that's the next wave," Mogentale told me. Nutrition studies agree: Fiber is the magic bullet for a happy, healthy gut.

Our food system makes it hard to stay away from processing. For my own diet, I choose to look for the layers in food production, which means how many sources for ingredients, and how many steps to creation. It took only one hand when I used this unofficial method on Mogentale's chips. When I finally tasted samples, I thought they were great. The square-shaped chips delivered super crunch, and I could eat a small handful and feel satiated. Nutritionally they were close to regular tortilla chips, but the fiber in Pulp Pantry's chips was double. I quickly fired off an email to Mogentale. "They're great!" I wrote. "The sea salt needs a hair more salt, and the barbecue flavor had too much seasoning." What can I say, I've spent a lifetime looking for a healthier snack.

The Outlook

The questions I have around upcycled products involve how challenging it is to uncover each step of the manufacturing process, including the origins of each ingredient in its later formulation. This conundrum applies to all processed foods. Technologies like blockchain could tease some of this out. But blockchain, a way to include a digital record at each step along the supply chain, is far from even early adoption. I'm also uncertain around the nutritional quality as ingredients are pressed, heated, and boiled. Did they retain their healthfulness?

Author and nutritionist Ginny Messina—online @TheVeganRD—promotes a whole-food, plant-based diet, but agrees that there is a place for processed foods. She's even a fan of the Impossible Whopper, which she considers fine as an occasional treat. One good example of beneficial processing Messina pointed to is when milk made from soybeans is stripped of most of its fiber. The result—blocks of tofu—is an excellent source of calcium. "Some processing can enhance the food or make it more digestible," she explained.

I also asked McHugh from the USDA whether processed foods can be healthy. "A lot of those components are heat stable, like fiber and polyphenols." Polyphenols are naturally occurring compounds found largely in fruits, vegetables, cereals, and beverages. There are strong suggestions that a diet high in polyphenols will provide a diet rich in antioxidants. But "every material is different," according to McHugh. The USDA claimed to "optimize" around nutrition, but that sounded like marketing-speak to me.

Many of us have olive oil in the kitchen, but we rarely consider its by-product, which is olive pomace. The olive oil industry probably doesn't want to hear this, but McHugh said that "some of the health-promoting compounds are found more in the pomace." McHugh is working with large producers to find uses for it. Unfortunately, capturing pomace requires restructuring the manufacturing process to ensure both quality and safety for human consumption—something every company in this chapter has had to resolve. Renewal Mill added equipment and additional manufacturing steps to Hodo Foods' tofu production line, which had to expand the footprint of its manufacturing facility to accommodate the changes; and ReGrained built their own plant in Berkeley to process its flour. According to McHugh, though: "Most companies aren't looking to reinvent their manufacturing line." Based on investment from bigger beer companies like Anheuser-Busch and Molson Coors, this sentiment could be shifting.

To rescue waste requires time and investment, neither of which are line items companies want to tack on. Turner Wyatt, president of Upcycled Food Association, sees the value for big companies to work closely with these upcycled startups because they can help them meet their new corporate responsibility agendas: "Companies are going to be under fire for the dates that they set their sustainability goals," he said. In order to meet their deadlines, many set for 2030, Big Food can align with these companies to get there faster. For Deutsch, who's also on a task force created by the association, one of the controversial issues is how companies show their environmental benefit. "Is it a reduction in GHG, a diversion in food waste, or is it enough to take it out of landfill?" he asked. Were I on the task force, I might propose that Big Foods' sustainability goals must also include reductions in manufacturing-related food waste, and the required use of upcycled ingredients. (In a perfect world there would also be broader guardrails on the wide reach of nutrient-poor snacks.)

In today's landscape, sustainability is good business. A 2017 study by the NRDC found there was an average fourteen-fold financial return on investment for companies that implement efforts at reducing food waste. But before they can get there, the small upcycling cohort needs to mature. Kurzrock of ReGrained elaborated on areas where upcycling has room for improvement. "It's diluted as a concept. It's underutilized and overlooked. There are flavor challenges in formulating it into food and/or flavor maskers." Finally, because upcycled ingredients tend to have less gluten and more fiber, foods that are formulated with them typically require another flour to make them more efficient. This means that the amount of upcycled ingredients is only 5 to 10 percent of the final product, sometimes less.

It may seem that the future of upcycling is assured, but our food system is rigid and reluctant to make change. Sarah Masoni, my Oregon State University food innovation expert, told me about the time she had a grant from the state to research the viability of reclaiming pomace—from grape skins and seeds—for food products. "We approached the wine industry in Oregon to ask them about it, and they laughed at us." Thinking this had to be an old story, I asked her what year the meeting happened. "In 2019," she said. "It's cheaper to throw it away than to use it."

But there's much to be gained from looking into the forgotten corners of our food system, and upcycled products continue to appear on the shelves. There are drinks and fruit snacks being made from cacao pulp, the white exterior around cacao pods; tortilla chips made from leftover corn germ; sparkling water made with the juice that comes from dehydrating fruit; banana bites made from overripe bananas picked on the farm; and coconut jerky made from the remnants of coconut water production. All manner of flours are being made from green banana peels, coffee cherries—the husk around a coffee bean that goes mostly unused—and even a new spice made from vegetable scraps. Ripple Foods has a grant from the Gates Foundation to work on low-cost formulations of its milk from the defatted meal of oilseed crops. (They're also working with McHugh and her team at the USDA.) The Berkeley company has already seen promising results from fava beans and canola seeds, and hopes to have it on supermarket shelves at a lower cost than its pea milk.

Price is the final hurdle for New Foods. Until these foods become more economical for a wider community, and are made in greater forms and flavors for a globally diverse audience, they will continue to be made for the elite who have money to spare, a curiosity for what's next, and a desire to wear the virtuous halo. For what it's worth, I'm guilty of supporting these innovations with my wallet and curiosity. I want to believe that good can come from change and I hate wasting food. But turning processed ingredients into processed foods gives me pause. What was added to make it palatable? Sugar, fat, salt, and flavor. Cheez-Its are delicious, but I don't need more tempting versions of them to exist in this world. Drexel's Deutsch was more pragmatic: "To the extent that the food system is producing varying [foods of] nutritional value, these upcycled products are also going to be of varying nutritional value."

Chapter 6

Plant-Based Burgers

Can Plants Replace Red Meat?

Just a Bite

In 2017, a few months after its launch, I ate my first Impossible burger at the San Francisco temple of meat called Cockscomb. In addition to curiosities like beef heart tartare and pigs' ears, chef Chris Cosentino—who was introduced to the "meat" by chef Traci Des Jardins after she was hired by Impossible to lead the culinary launch*—served up the Cockscomb Impossible burger, layered with lettuce, Dijon mustard, Gruyère cheese, caramelized onions, and bread-and-butter pickles. Planted into the pillowy bun was a toothpick waving an Impossible flag. It was massive; if there was a suggested serving size, this was twice that. I bit into it, tearing into the brioche bun and inch-thick patty, uncovering a lushly textured, pink-hued center. It was exactly as they had marketed: a burger that bleeds, albeit one made entirely from plants. Save for the

* *Impossible: The Cookbook* came out in July 2020 from Des Jardins; it was self-published by Impossible.

fact that it was more crumbly than a ground-beef patty, it fooled me into believing it was actually from a cow.

Sitting across from me was Jessica Appelgren, Impossible Foods' friendly and chatty vice president of communications. She admitted that the formula needed work. Impossible is one of many companies trying to replace the 2.4 hamburgers an average American consumes in a week with a greener, more ethical alternative. "We want to be habitual status," she chirped. I imagined she meant "habitual" in the way that fast food became a habit—a quick option that went from conscious choice to an automatic routine based on external cues. Fast food is certainly one of the sectors to blame for the decline in the world's health. The Impossible marketing message is not to think about what you're putting in your body, only that what you're eating is better for the planet.

To get to the bottom of how this patty made from plants could taste, feel, and bleed like real meat, I drove to a sixty-seven-thousand-square-foot warehouse located in an industrial section of Oakland, California. Opened in 2017 in a former dessert factory that baked up cakes and cupcakes, it was Impossible Foods' first manufacturing plant. Later, when the company was struggling to keep up with orders, it signed a deal with OSI Group, a food producer based in Illinois that operates sixty-five plants in seventeen countries. In another sign of progress, Nick Halla, employee number one, moved to Hong Kong and is working on the company's launch into Asia.

Appelgren met me in the lobby and took me upstairs to an empty conference room. Almost no one was in the building. Most of Impossible's staff works in Redwood City at an unremarkable office park in between two massive recycling centers. While we waited, she offered me a cup of strong coffee. We were soon joined by Chris Gregg, Impossible's then chief supply chain officer, and plant manager Julien Grascoeur, a tall Frenchman who seemed eager to show off the almost brand-new assembly line. After some small talk in the conference room, we suited up in white lab jackets and plastic eyewear, and headed inside the factory.

The cement floor was spotless. Yellow lines demarcated areas of work, and safety signs and hazard warnings were posted at eye level. Stacked high on tall metal shelves in a side room off the main factory floor were cardboard boxes and tote bags of ingredients. I glanced at them, wondering how dry goods became a beef-like burger that could fool meat lovers. In the ultimate of cloak and dagger, Appelgren, who was beside me at every step of my tour, told me not to read the labels.

Inside the main production area, paddles on stainless-steel machines churned an amalgam of plants. My nose wrinkled. The place stank—of what, I wasn't sure. The Impossible burger is a mishmash of seventeen ingredients, including heavily processed items like textured soy and potato protein plus pseudo-healthy additives like coconut and sunflower oil, riboflavin, and zinc. What exactly was I smelling? Was it the heating mechanism? Or maybe it was the process of turning the bits and pieces into plant-y mush? I asked questions, but many went unanswered because the information was proprietary. Impossible has filed around 140 patents on everything from extracting and purifying proteins, to soy-based cheese, ground meat replicas, and genetically engineering methylotrophic yeast, which is the mastermind behind the heme molecule. More on heme (the word rhymes with "team") in a moment.

Touring the cavernous building, I couldn't help but think of "truth in advertising" laws. The Impossible burger did "bleed," and there was a dark red river running below the equipment to prove it. No doubt the smell came from that. Later I realized the smell most likely came from heme, which enables the Impossible burger to have what's called the Maillard reaction—a caramelization that goes from red to brown—when it's cooked. The operation reminded me of meat processing plants I'd seen before—the scent, the mess, the chill—except for one key difference: no actual blood was shed.

When Burgers Became King

What was once a culinary category wasteland—veggie burgers, now marketed by the trendier name "plant-based burgers"—has captured the attention of investors ranging from Bill Gates to Shaquille O'Neal. There's a reason they're throwing their money at these once undistinguished and unpopular patties. One out of four US consumers decreased their meat intake in 2019. In a report from the Plant Based Foods Association (PBFA),* the plant-based meat category grew by 29 percent in the past two years, to $5 billion. More alarming for the traditional meat category, refrigerated plant-based meat (including Impossible and Beyond) grew 37 percent, while conventional meat grew by only 2 percent.

This sounds promising, doesn't it? However, worldwide, plant-based meat sales are valued at $12.85 billion. Traditional meat is valued at $2 trillion. Because of the near-constant headlines, we think that plant-based is killing it when in reality it has captured just .6 percent of the market. While plant-based meat saw phenomenal traction in supermarkets due to changes in our shopping habits during the pandemic, it will take immense efforts to compete with meat.

What Impossible and others are doing has been attempted before, with less success. In 1896, the Seventh-day Adventists, a conservative Protestant denomination who believe that the Bible promotes a vegetarian diet, introduced a meat substitute they called protose. Made from soy, peanuts, and wheat gluten, it was ground into a thick paste, mixed with water, thickened with flour, steamed, and sterilized. This process isn't too dissimilar from what startups are doing today.

Protose was sold in tin cans and distributed by Battle Creek Food Company. The company was founded by John H. Kellogg, brother of W.K., of the cereal fortune. In 1944, Clementine Paddleford, a syndicated food writer, wrote about one of these faux-meat contenders in the *Baltimore Sun*:

> Beanburger is a meatless meat, packaged as a powder, to whip up with water, form into patties and fry like hamburger. A

* Report based on data from SPINS released March 3, 2020, and done in collaboration with the Good Food Institute.

protein derivative plus a nice blending of spices give the meaty flavor to this mixture of soybean grits, wheat flour, cracker meal and dehydrated onions. For a tastier treat, use it as a stretcher, half and half with ground round steak.

In 1947, after the Second World War and at the end of meat rationing, the Waldorf Astoria Hotel in New York served an hors d'oeuvre loaf made from protose—most likely shaped like a meatloaf—referring to it as "an unusual combination very popular with those of sophisticated tastes." How this was seasoned is anyone's guess.

Over the next few decades there were other attempts made to offer credible alternatives to the beef patty, but the substitutes were bland and mushy, even when put in buns. More closely aligned to overcooked vegetables than legitimate meat, these products failed to convince anyone, and adoption beyond natural food stores didn't really matter either. In that era, the oil-derived fertilizers that fueled intensive farming needed to grow food to feed cows, and the land needed to farm was not a concern to most. Lappé was the exception. In 1971, when she realized that half of all harvests went to feed livestock, she declared that "a meat centric diet was like driving a Cadillac." It was sucking all of our resources into a low-return product that amplified our economic divides. "Those who want the grain and need the grain cannot afford to buy it, so livestock get it." In the United States alone, 56 million acres of land are used to grow feed for animals, and 4 million acres are used to grow plants for humans.

In spite of these new tempting analogues, demand for animal meat has never been higher. In a blog post on Medium, Paul Shapiro, CEO of Better Meat Co., wrote that while the pandemic has helped to drive record adoption of these new plant-based brands, 99% of the fresh and frozen meat sold at grocery stores continues to come from conventional animal meat. You can do the math.

Burgers are so ubiquitous, so quintessentially American, that it's hard to remember that our national love affair didn't begin until 1955, when McDonald's opened its doors. Still the largest fast-food chain based on revenue, and operating in 119 countries, McDonald's no longer tracks how many burgers it sells annually, but a quote on the Internet from an old McDonald's training manual suggested the burger giant sold "more than seventy-five hamburgers per second, of every minute, of every hour, of every day of the year." It's safe to say the number of burgers is in the many, many billions annually. In an episode of food podcast *Gastropod*, I learned that the National Center for Health Statistics reported that more than 36 percent of the world eats fast food on a daily basis.

With that many burgers being eaten, it's becoming harder to ignore that our American diet has colonized the world. Essentially, we are to blame for the twin downward spirals of climate change and poor health. That realization has driven a new breed of entrepreneurs back to the burger. Researchers estimate that the market for meat substitutes—often called analogues—is growing rapidly. Armed with the latest in food science, Impossible Foods, and its closest competitor, Beyond Meat, introduced slightly different plant-based burgers with the goal of capturing the growing global demand for beef, and to give Americans, newly passionate about eating plant-based, a serving of "meat" they wouldn't feel guilty eating.

Engineering a Better Burger

To date, these two companies have raked in enough funding to run a small country for a year. As of August 2020, Impossible had raised $1.5 billion. Prior to its initial public offering (IPO), or stock sale, Beyond Meat had raised at least $122 million. At its IPO in May 2019, Beyond Meat raised an additional $240 million, despite warnings that the company may never make a profit. Why are investors betting that these new iterations of the meatless burger can succeed where others have failed? And what does our obsession with burgers in all their forms tell us about ourselves?

The founders of these two burger behemoths, both vegans, say they're doing it to combat climate change, which they say is directly linked to our reliance on animals for food. Pat Brown, who started Impossible in 2011, is a former Stanford biochemist and a world-renowned geneticist with several awards and honors. His invention, the DNA microarray, which is used to study and analyze genes, is still in use today. So, he's brilliant. When I went to meet him in his office on October 26, 2017, I was ill prepared for the debate, which is what it felt like.

We met in the Impossible headquarters in Redwood City, near San Francisco. Brown wore the casual dress of a Silicon Valley engineer—running shoes and a hoodie. Next to us was Appelgren, who typed our conversation onto her laptop, and chief communication officer Rachel Konrad, who would glance at Brown anytime he began to share something off the record, and stop him.

It was during a sabbatical from Stanford when Brown decided he wanted to be remembered for doing something radical: ending our reliance on eating animals. The problem, Brown said, is that we conflate meat with the animals that produce them. In a post on *Medium* he wrote: "Until today, the only technology we've known that can turn plants into meat has been animals." An efficient man, Brown is the epitome of a scientist—hard facts, data sets, evidence-based. But then his passion takes over and you sense something of a zealot in his mission to feed the world plants that taste like "beef."

Early iterations of the Impossible burger tasted like "rancid polenta," according to one team member. "What we have now is better," said Brown. "What we have in six months will be [even] better." When the company ran blind taste tests in the Midwest, people were given two things that looked like meat, the Impossible burger and a typical beef burger, called an 80/20 for 80 percent lean ground beef and 20 percent fat. At the time, half of the tasters preferred the plant version, he said. At 46 percent, it was almost half, according to an Impossible-commissioned study, but since the launch into fast food, it's likely that number has well surpassed half.

For some chains, the Impossible burger beats out traditional choices. At Umami Burger, with more than twenty-two locations around the world, one of the top five menu items in gross sales was the Impossible burger, which features two Impossible patties, grilled onions, vegan American cheese, miso mustard, Ooh-Mami sauce, dill pickles, lettuce, and tomato.

Nutrition-check: Those two patties alone contain 480 calories, 16 grams of saturated fat, and 38 grams of protein, not counting the condiments, bun, or fries.

It's true these burgers appear to be better for the planet. Impossible claims its burger uses 95 percent less land and 25 percent less water and releases 89 percent less greenhouse gases into the atmosphere when compared with industrial beef production. Even McDonald's is feeling the pressure to improve sustainability and has vowed to eliminate deforestation from its beef supply chain by 2020. The fast-food giant, the single largest buyer of industrial beef, claims it's also reviewing antibiotic use in its meat by 2020, but I could find no data supporting these two claims. Let's put a pin in whether Impossible can scale up to supply burgers to McDonald's and hang on to their better-than environmental numbers.

Recently, nutrition scientists have been urging the world to be more mindful of levels of food processing. One method that is getting traction is the NOVA scale, which puts food into four distinct buckets: unprocessed or minimally processed, like seeds, fruit, eggs, milk, fungi, and algae; processed culinary ingredients like salt, sugar, olive oil, and vinegar; processed foods like bread, cheese, and smoked meats; and finally ultra-processed foods and drinks like soda, ice cream, burgers, and instant soup. A burger from Impossible falls in the last category. The Brazilian researchers behind NOVA write: "Common attributes of ultra-processed products are hyper-palatability, sophisticated and attractive packaging, multimedia and other aggressive marketing to children and adolescents, health claims, high profitability, and branding and ownership by transnational corporations." It's also because it has seventeen ingredients in total, each made by a different company, and some that are "additives whose purpose is to imitate sensory qualities."

Outside of its spot in the ultra-processed category, which no one in the company relishes, the ingredient most often worried over is heme, which is a genetically modified version of iron. It's an integral part of many proteins and a fundamental part of animal meat. This includes human muscle tissue. In meat it's myoglobin; in plants it's leghemoglobin and it's referred to as "non-heme iron." Impossible's version (nicknamed RUBIA) originates from the root nodules of soybean plants and is made by genetically engineering yeast to produce mass quantities of soy leghemoglobin. Impossible uses the shorthand "heme." Like blood in animals, heme is red. That was the red river I saw running on the factory floor. It smelled like blood because it kind of *was* blood. Just not from an animal.

We've been eating food that owes its existence to people in lab coats for decades. The current trend is no different, but do we really understand what Impossible Foods is selling?

Dr. Dean Ornish, a professor of medicine at UCSF and an early proponent that a whole-foods vegetarian diet and other lifestyle changes could reverse coronary heart disease, had concerns. "It needs to be studied more," he told me. "Heme iron from red meat is much more readily absorbed [by our cells] than heme from vegetables, causes oxidative stress, and is linked with an increased risk of coronary heart disease, stroke, and other chronic diseases." This is one of the reasons most doctors frown at meat-centric diets. A meta-analysis of prospective studies from 2014 found participants with higher heme iron intake, which comes only from meat, poultry, and fish, had a 31 percent increased risk of coronary heart disease, compared with those with lower heme iron intake. My question to Impossible, and which Dr. Ornish could not answer, is whether Impossible's heme is equally bad for heart disease in humans.

Nonetheless, Dr. Ornish, author of seven books on health and wellness, is happy to see these new products on the market. "Anything that helps more people eat a plant-based diet is great," he said. But, he posited, if the jury is still out on the potential risk of the heme in the Impossible burger, and you can make a good-tasting plant-based burger without it—like Beyond—why not go that route until definitive research is completed?

Unlike the Impossible burger, Beyond's version is made without heme. When Ethan Brown told me he avoided using heme in his formulation, he quipped that Impossible's embrace of the GM ingredient could be "brilliant or its Achilles' heel." Instead, the Beyond burger is made primarily of pea protein, although the company says it's protein agnostic and is trying other sources like mung, canola, and fava. Despite Beyond's early use of pro athletes for its influencer marketing campaigns, it, too, falls in the ultra-processed category.

In order to quell consumers' fears about a genetically modified ingredient, Impossible applied to the FDA for a GRAS ("generally recognized as safe") determination on its version of heme, something it was not required to do. The company wanted a GRAS so that it could point to it as proof of safety. The FDA initially rejected the petition, noting that "soybean root is not a commonly consumed human food," and that Impossible did "not establish safety for consumption." Digging deep, Impossible ran an additional study where scientists fed soy leghemoglobin to rats every day for a month at two hundred times what the average American consumes daily from ground beef. In a 1,066-page response to the FDA, Impossible reported that it found no adverse effects. But as I've noted before in this book, this was a study performed by Impossible, *not* an independent analysis. "One of our core principles is that we are never going to sell food for consumers that is not better than what it replaces," Brown told me in our interview.

Why fight so hard for one small ingredient? Because Impossible has a whole product line in development that depends on heme: eggs, chicken, pork, and fish. Barring procedures regulating the food industry, the company is hopeful to expand into mainland China. Although industry experts suggest it may be a challenge because of the company's genetically modified version of heme.

In the United States, there's almost no stone unturned by the billion-dollar startup. Where once Impossible only sold to chefs and food service, it now sells in supermarkets and fast-food chains, and on its own e-commerce website. In late 2018, the FDA finally granted Impossible GRAS certification for the Impossible burger, lifting the barriers to consumer acceptance. It took longer to get approval to use it as a coloring additive—heme is what makes the burgers look pink—but that was later approved* as well.

During my research on protein isolates, I learned that the original Ralston Purina soy manufacturing plant was acquired by DuPont, a well-known chemical company that is best known for dumping toxins in our waterways. DuPont produces around 36 percent of genetically engineered soybean crops. Somehow these two facts made me think that the company was working with Impossible. When I attended a panel at a food tech conference in Brooklyn, a scientist from DuPont illustrated how enzymes were used in food manufacturing. I perked up! During the audience Q and A, I asked what it was like to work with Impossible on creating novel ingredients like heme. He answered in anodyne terms, saying that it was an interesting collaboration. Offstage, I tracked him down to inquire where they were making the ingredient. He shared that DuPont was making the heme at a facility in Mexico.

According to the Impossible founder, beef flavor depends on heme. While only .02 percent of each burger is heme, staffers report that without it, the burger tastes like a crab cake. Crab sounded like a reach, but I never had the opportunity to try an Impossible burger without its heme. It's a catalyst, there to spark amino acids, sugars, and fatty acids into something our palates recognize as meat. It's also there to set Impossible apart from the competition and to lure investors in with its unique know-how. If plant-based protein can't taste like meat without heme, how can any other company do without it?

The Brown of Beyond

Compared with Impossible Foods' high-tech Stanford allure, Beyond Meat feels like something that began in a garage down the street. Its headquarters are located in El Segundo, California, a coastal town better

* To approve the color, the FDA relied mostly on a short twenty-eight-day study, which does not translate to long-term safety. Impossible's petition for its color additive was filed on November 5, 2018, and the FDA approved it on July 31, 2019.

known for its refineries and a song by A Tribe Called Quest. Beyond is the startup with the second-biggest IPO valuation in the last decade, although Airbnb and DoorDash came close in 2020.

Ethan Brown (no relation to Pat from Impossible) is a towering ex-jock who spent weekends on his family's dairy farm in Maryland. Armed with an MBA from Columbia University, and some work experience in clean energy, Brown, also a vegan, wanted to prove that "you didn't need an animal to produce meat." He combed through research papers to find a technology that might work for his mission, and launched his company—initially called Savage River, after his family's farm—in 2009.

In the early days, Brown worked with Dr. Martin Lo, a food scientist and former professor from the University of Maryland, his alma mater. First up was vegetarian "chicken." The pair hoped to replicate the stringy protein fibers that ribbon through chicken meat. But it wasn't easy. In an alumni magazine, Lo reported: "The first generation of the product was like a broken tire torn apart."

When he finally had a product he felt good about, Brown spent hours on his feet handing out samples in grocery stores throughout the Midwest. "Women would come up to me and ask, 'How do I get my husband to eat this?' " Brown told me in one of our many interviews, both in person and on the phone. He also fed it to his family, including his teenage son, who, he said, eats multiple Beyond burgers every week.

Beyond burgers are produced in a multi-step process of heating, cooling, and pressure, which re-stiches the plant fibers together. It takes two minutes to make a burger this way, or fourteen months to raise an animal. This length of time is crucial to these founders' save-the-environment platform. They like to point out how inefficient animals are at converting energy into human calories—it takes 23 calories of input for a cow to convert into 1 human calorie. (For reference, chicken is the most efficient at 9 calories in for 1 calorie out.)

One big problem with the efficiency argument is that it works against plants. Spinach takes six weeks to grow. A tomato takes three months.

Are we sleepwalking toward a future in which future foods must be harvestable faster than their traditional counterparts? Will it ever get to the point that someday we are led to reject any food that takes longer than a few minutes to grow or make?

The plant-based debate is sure to rage on, with a range of companies churning out their own iterations of burgers and nuggets, but calling these things plants is like calling a Slim Jim meat. While I make this comparison a bit off the cuff, when I looked up what goes into a Slim Jim jerky stick I was pretty horrified. In an issue of *Wired* from 2009, the tech magazine reported that "poultry scraps are pressed mechanically through a sieve that extrudes the meat as a bright pink paste and leaves the bones behind (most of the time)." That's not far off from what the ground "beef" looks like. Both products contain dextrose, which is an antimicrobial ingredient and keeps bacteria "asleep." Where a Slim Jim includes sodium nitrate to keep the meat from turning brown, the Impossible burger includes tocopherols in addition to heme, which could be helping its pink-when-cooked color retention.

Are these different products? Yes. Are they similarly made? Yes.

How Do They Taste?

By fall last year I'd eaten each company's burger a handful of times, including both companies' breakfast sausage. I'd eaten the Beyond burger at a crowded bar alongside Ethan Brown. It was cut in half so I could see the brown edges and pinkish-hued center. It captivated me as it sizzled on the griddle. Neurons in my brain and taste receptors in my tongue told me it was tasty, but my brain and belly fight over this constantly. Do I like it because it's decadent and, like all burgers, a perfect foil for the accompanying salty, sweet, and umami condiments? Foods engineered to be delicious usually come with a cautionary tale. Sure, Doritos are good. Twinkies, too. But should I be eating them daily or even weekly? I doubt it.

On the topic of burgers, Ginny Messina, the VeganRD, mentioned she had eaten an Impossible Whopper the day before we talked. I was surprised, but the plant-based advocate thought nothing of it. She loved her Whopper, but declared it fast food, not health food. "I think they are really good and fun, but it's not my whole diet, it's a treat," she said. The problem? To pay back investors, these companies need their burgers to be sold at frequencies well beyond the occasional treat.

When I cooked up one of Beyond's burgers at home in a pan, it left behind a thickly oiled, fried food odor that took days to dissipate. I ate it dipped in liberal amounts of Dijon mustard and spicy ketchup. As a vehicle for my beloved condiments, it did the trick. But it was heavy in my gut, not unlike real meat. When I ordered the burger from Impossible at Gott's Roadside, in Napa, California. I added ketchup and mustard to the lettuce, tomato, and American cheese already present. Then I took a giant bite. Eyes open or closed, it was a convincing burger.

What impressed me about both burgers was the texture—the tug and chew of believable fibers and fat—something that many people identify as what we crave most from meat. Instead of the flavorless, mushy veggie burgers of old, these new plant burgers are convincing replacements. They're salty and packed with flavor. Coconut oil matches animal fats in a satisfying richness, and we are rewarded with a deliciousness that excites our brains, which beg for another hit.

Virtually every New Food company nods to both Impossible and Beyond for paving the way for the exciting increase in plant-based foods, and I am eager to see how this develops. Right now, we have mimicry of traditional animal products—beef, chicken, fish—and we have a very American, white male framework—the burger. Will there be a future where we see foods we can't even begin to imagine? Will there be new players in the field that offer us regionally focused foods that speak to different cultures and ethnicities, like the many meat replacements that already exist in Asia? I hope so.

The Path Forward

In February 2019, Impossible changed its burger recipe, swapping out the wheat for soy in order to make it gluten-free. Their marketing referred to it as Impossible 2.0, like it was a software release. In January 2020,

Impossible launched its newest product, pork sausage, at the Consumer Electronics Show in Las Vegas. For the "pork," *Business Insider* reported that Impossible had "tweaked its heme formula to better replicate the consistency of pork rather than red meat." So now Impossible's heme can color the burger from pink to brown, it can make the burger taste like beef, *and* it can improve the consistency? This seemed too good to be true.

Back in the lab, the team is now working on the flavor of steak.

Food science has long been involved in crafting our dinners, but today's food startups retain an enormous amount of intellectual property from the years of lab work that they've invested. They worry about getting patents. I worry that they don't share because it's what investors demand—proprietary products, even when it's as basic as everyday sustenance. My questions to Impossible's Pat Brown, the same questions I asked every other company in this book, was whether today's call for greater transparency is being heeded or whether new technology is shrouding the food manufacturing process? My need for specific details, answers that I thought would assure me that they were looking out for my own best interests, were very hard to come by.

Brown assured me it wasn't what he wanted. "Consumers have to know what they are getting in a product. The stuff that is sensitive is not something that will move the needle on judgment." Later, sitting in Brown's office, he promised me that he would share the company's secrets once it's awarded the appropriate patents. The US patent process takes two or more years, and Impossible has 139, or more, in the pipeline.

These assurances are easy to make, but I have many questions. Emails I sent to the Impossible communications team were met with resistance. They accused me of digging for dirt, and bait-and-switch. The chief communications officer wrote to me, "I'm starting to doubt whether our data will be presented fairly and in the most appropriate context." This was based on my asking questions inquiring about ingredients and the production process—things I would do with any topic I aim to cover. Sure, it's annoying, but it's not out of line. Eventually they went so far as to assume that because I asked about the temperature in the factory, I must be planning to construct my own life cycle analysis (LCA) on the burger, a Herculean task even for a scientist with access to sophisticated lab equipment and graduate students to crunch the numbers, let alone a solitary journalist.

Before all of this, when I sat with Brown in his tiny conference room alongside his communications staff, he assured me, "We want to be transparent about how we make our heme." Yet it's still a trade secret, something Brown promised not to rely upon. Investors don't make their investment back if he shares his science. Brown's last vow to me that day, "never to sell food to consumers that isn't better than what it replaces," is simply another impossibility. Until he comes up with a reliable nutritional study, the jury's still out.

Chapter 7

Vertical Farms

Can Premium Greens Picked by Robots Feed the World?

Your Leafy Greens Are Created by Algorithms

"You never really hear anyone saying, 'I love the way kale tastes, I love to chew kale, or I love how bitter it is,'" Alina Zolotareva explained when I visited her at AeroFarms, a seventy-thousand-square-foot vertical farm in Newark, New Jersey. Worse, she said, they'd been hearing from their customers that kale was "difficult" to work with. "Chefs have to de-rib it, they need to chop it, marinate it and let it sit, or acid massage it. It takes so much time, so many labor hours, and it's really, really difficult."

She's right, but none of that should matter. Kale is a king among salad greens. By eating one cup of kale a day, you're getting fiber, antioxidants—especially alpha-lipoic acid—calcium, potassium, vitamin K, vitamin C, vitamin B_6, and iron. You even get 3 grams of protein. For these qualities alone, we should be mainlining kale.

Actually, we are eating a lot of it. In addition to countless bowl-based chains, the hardy green appears in a southwestern salad at McDonald's, a kale crunch side salad at Chick-fil-A, and a Greek salad at Panera. But vertical farms like AeroFarms—where everything is grown inside giant buildings, usually in urban areas—are not the suppliers.

Vertical farms don't talk about wanting to land the McDonald's account. Instead, they talk of their products' safety—untouched mostly by human hands. Also that it's more delicious, and fresher, because it has fewer miles to travel. If vertical farms succeed, we'll eat entirely new versions of fruits and vegetables: sweeter kale that is delicate and light, arugula that is spicy but not too spicy, and watercress that doesn't rot in your fridge the day after you buy it. When people talk about the diet of the future becoming more personalized, it's not too outlandish to think that farms might one day deliver greens tailored to our personal needs. Messaging on the bag could tout: *Higher Potassium to Lower Your Blood Pressure!*

There's a flip side to vertical farms, though. Are foods grown in these highly specialized environments as nutritious as produce grown on a traditional farm? This question speaks to nutrient density, as well as to hidden factors that we have yet to fully explore. Specifically: Is this new breed of soil-less kale as beneficial for our bodies as the traditional dirt counterpart? Crop science is just beginning to research what the microbiome of soil means to human nutrition, as well as the crucial interaction between microbes found in soil and the roots of plants. If vertical farms become normalized for fresh food production, what interactions are lost that are essential, or even just helpful, to our survival?

Another point to consider is that when a pathogen gets in to an indoor environment, it's really hard to control. While we've seen frequent recalls of produce from the fields, and the resulting infections that arise from it (especially romaine lettuce after *E. coli* bacteria contamination), we have little inkling what could occur if a vertical farm experienced a large-scale outbreak and had to recall hundreds of thousands of units. (Although it's possible we can assume that nothing leaves the building without thorough testing, we can also assume there are dozens, and maybe many more, factors to watch over.) On the sustainability side, how green are these giant factories that rely heavily on the grid (only sometimes using renewable energy) and on expensive man-made fertilizers in order to churn out premium greens in an endless parade of single-use plastic tubs?

The Way Back Machine

Vertical farms didn't launch into our world overnight. One of the first people to demand his plants grow outside of their normal seasons was the Roman emperor Tiberius Caesar, who in the second century AD was told by his doctors to eat a cucumber a day to "cure an imperial illness." At 96 percent water, the unassuming cucurbit could have been prescribed for its hydration, but it also has fiber—including pectin, which promotes regular bowel function. Regardless of the reason, growing cucumbers year-round for the emperor required gardening innovations that included protected plant beds that could be moved indoors and out, depending on the weather, and raising the temperature during cold months by adding manure (a natural fertilizer) that could increase heat on the plants, forcing them to grow earlier, and more frequently.

A somewhat extreme step for one person, these cultivar methods didn't pick up until the early thirteenth century, when they began spreading throughout Europe and Asia to grow delicate treasures that explorers brought back home, including oranges, lemons, pomegranates, myrtle, and oleander. In Italy, these buildings were called *giardini botanici*. Fully enclosed from the elements, a typical greenhouse traps heat inside or can be heated mechanically. Yet, they still depend on soil and sun.

Greenhouses proliferated, but it wasn't until a scientist at UC Berkeley, William Frederick Gericke, published his findings in the *American Journal of Botany* in 1929 that it began to gain traction. Initially, Gericke focused on wheat, which he was able to grow faster by placing it in tubs with a constant supply of water and nutrients, and near-constant "sun": argon-filled lamps pointed at the stalks for sixteen hours a day. Eventually, he could also grow tomatoes and other crops this way. His article, titled "Aquaculture: A Means of Crop-production," showcased his initial name for the niche farming method. When he learned that "aquaponics" had been used in fishing since the mid-1800s, he turned to one that a Berkeley colleague had suggested: hydroponics.

Soon after Gericke's article, futuristic agriculture began popping up at World's Fairs. In 1939, in Flushing Meadows, New York, organizations presented their visions for the future, a world where food production was automated and no hands got dirty. To show off its ketchup, Heinz built a hydroponic garden with ten-foot-high tomato vines. It was called the "Garden of the Future." Borden (a now defunct food company) demonstrated its automatic cow-milking Rotolactor. The USDA, on the heels of the Great Depression, focused on promoting food production with a banner atop its nutrition exhibit that read: Men=Chemicals=Food. Not too far behind in touting its science savvy, DuPont called its exhibit "Wonder World of Chemistry." As a form of entertainment, these fairs are unequaled by any event today. Almost forty-five million people attended the World's Fair at Flushing Meadows. My grandmother was one of them.

In a frame inside my home is a black-and-white photo of my grandmother sitting next to her father outside the fair. She's wearing a floral-patterned dress, knee socks, and white sneakers. She looks so young, not at all as I remembered her—wrinkled with a bun of gray hair atop her head. I wonder if she saw any of these futuristic gardens, or if she stood in line to buy samples of General Foods' new frozen foods. After my grandfather died at an early age (he also had type 1 diabetes), she went back to school, and then to work as a landscape gardener. In the backyard at her home in Van Nuys, California, she tended beautiful, flavorful tomato plants.

In the 1970s, indoor farming gained ground as a way to combat future food shortages that experts predicted, a cautionary tale that reappeared frequently in mainstream media. Today, the same fears are being ignited with a similar question: "How will we feed nine billion by 2050?" A USDA report from the seventies ushered in a formalized name for this futuristic vision: "controlled environment agriculture," or CEA. The report's author, agricultural economist Dana Dalrymple, was interested in the prospects of greenhouse food production. He outlined the evolution of the industry and how a CEA might operate dialing up or down various grow settings, including for temperature, light, air flow and composition, and root medium. "Generally, the higher the temperature, assuming carbon dioxide and light are abundant, the faster photosynthesis takes place," he wrote.

While today's CEAs might be more advanced than Dalrymple ever imagined, the basics are the same. Seeds are planted in a potting medium, anything from coconut husks to rice hulls to hemp; even synthetic materials are used. After being planted, they're germinated in warm, enclosed environments, and set under grow lights for extended periods of time. Compared with sunshine, which is inconsistent in duration and intensity, indoor lights create a constant cycle for growth to occur, and specific spectrums can be dialed up or down depending on the seed. Along with frequent light, plants are fed a nutrient-rich solution that mimics what it might get outdoors from soil or fertilizers. Different crops are fed different mixtures and amounts of nitrogen, phosphorus, and potassium. Because an indoor environment can accelerate growth, indoor farms harvest crops more frequently than traditional farms, and the foods grown are (generally) more consistent in yield and flavor. Some indoor farms report being able to grow 350 times as much produce as a traditional soil-based farm—a number that is sure to attract investors. Maybe it won't come as a surprise to learn that cucumbers, along with tomatoes and herbs, are one of the top three crops grown in greenhouses.

Cheap lights are the biggest reason why momentum for building CEAs is swelling. In 2010, the sector expanded rapidly because of improvements in lighting technology and drops in equipment cost. This encouraged tech entrepreneurs to jump into farming, a sector not known for quick and easy financial returns. Robots, artificial intelligence, computer vision, and the promise of feeding the world piqued their interest. But it remains to be seen whether CEAs can deliver improved food safety, shelf life, and quality to everybody, not just the well-off.

LED lights have allowed vertical farms to proliferate, but they're still the biggest operational cost for two reasons: energy to run the lights, which are on almost constantly, and energy to cool the room down, because of the heat produced by the lights. As they grow, plants "respirate," or breathe, which means indoor farms need to deal with the humidity produced inside. Rooms stacked high with plants packed in snugly are complicated to control for climate, so expensive HVAC (heating, ventilation, and air-conditioning) systems are needed to circulate air and cool things down. In a 2014 report from Purdue University, Victor Mendez Perez hypothesized that if the US agriculture industry followed a vertical approach, the electricity required for lighting would be eight times the amount generated by all power plants annually in the United States. On the other hand, vertical farms use 70 to 80 percent less water compared with conventional farming techniques.

Because indoor farms are dependent on reliable cheap energy, researchers doubt that entrepreneurs have found the right form and function to export around the world—is it in local shipping containers, empty city buildings, or massive warehouses? One frontier that makes perfect sense is space, and NASA has been sending seedlings up since 2001. Using a simple grow box with automatically controlled water and light levels, NASA has performed more than twenty separate agriculture experiments inside the International Space Station, including growing a lesser-known Japanese lettuce called mizuna.

In 2014, astronauts grew red romaine lettuce. It was frozen and sent back to Earth for testing. In 2015, astronauts were allowed to harvest and eat the romaine. Spoiler alert: No negative health effects were reported. In 2020, a study was released on the safety of space-grown lettuce in *Frontiers in Plant Science*. It was found to be free of disease-causing microbes, as nutritious as those grown on Earth, and safe to eat despite being grown at lower gravity and with more radiation. Despite NASA's interest in this easily grown crop, field-grown romaine has been recalled for contamination so many times that vertical farms are steering clear of growing it.

While Dalrymple's report was written almost fifty years ago, his outlook on the sector could have been written today:

> The possibilities for environmental control have indeed placed the upper levels of greenhouse food production on a par with industry. But this does not mean that the problems have been removed; far from it. While some of the uncertainties associated with weather may be lessened, they have only been replaced with the increased economic uncertainties and difficulties associated with high overhead, heavier operating costs, and a highly volatile market. The problems are no less, just slightly different.

With continued investments from industry titans, CEA farms appear to be here permanently, but financial growth is slow. Statista reported the vertical farm markets value at $4.4 billion in 2019. In 2025, the global statistics provider estimated it to reach $15.7 billion coming from an increased demand for organic food—a label that is up for dispute from traditional farmers—and population growth in dense, urban cities. Along with critical questions left unresolved, the industry is seeing turnover. Many farms can't make the financials work. In court data from 2019, Reuters reported that 595 CEA farms filed for Chapter 12—a form of bankruptcy. Until more farms move beyond small scale, it will remain to be seen what the right fit for the world is: massive, medium, or small in scale; and, more important, whether the work done here can eventually reach the people who need it most—those who are food insecure, living where access to fresh food is limited, or in communities with little arable land.

It's Aeroponic, not Hydroponic

Startups don't usually have $238 million in funding, so where I was sitting shouldn't have felt like a startup. But the walls were brick, the layout was open, and there was free food in the kitchen—all that was missing were staple startup furniture: beanbag chairs and a Foosball table. I was there to sample the greens, the only thing AeroFarms sells so far, despite having been in business since 2011. After the baby kale, which had a little slip of a water-filled stem and a small, slightly curled leaf, I tried Ethiopian kale, which had a larger stem, a darker green leaf, and a backward scoop. Where the featherlight baby kale tasted sweet, the Ethiopian kale had savory notes with a slight peppery finish. Baby pak choi (Chinese cabbage) was sweet and watery. Delicate arugula, made for mass appeal, didn't zing me with its usual pleasurable bite.

Plant breeding, which typically takes decades, is happening full tilt indoors at AeroFarms. According to CEO David Rosenberg, the team has tested almost one thousand varieties in the last year alone. In addition to her role in marketing, Zolotareva, who remarked on kale earlier in this chapter, is a registered dietician and a supertaster, which means her taste buds are more receptive to bitter flavors. Passing around a slim two-ounce tub of microgreens made especially for Whole Foods, she said that they were "super nutrient dense." The quicker you can get them from harvest, the better. I pinched up the delicate clover-shaped leaves and tipped my head back to catch them in my mouth. On its own, the flavor was subtle. They were, as a friend referred to them, tweezer food—delicate items placed at the last minute atop an expensive plate of food.

The team worked to find a variety they could grow quickly and harvest easily. After only fourteen days the fully grown plants are put on a belt and fed into a cutting machine set to automatically chop them down to a few inches. Next, workers place the plants on a chilled conveyor belt that slows down respiration, giving the greens a longer shelf life. Finally, the greens are hand-packed into plastic tubs. Before Whole Foods (owned by Amazon) would sign off on the product, its buyers asked for more purple. In response, AeroFarms added red cabbage sprouts, which are actually more purple than red. In our Instagram-ready world, more vibrant color equates to better sales.

AeroFarms grows aeroponically, which means the seedlings are not planted in soil, or even a potting medium, but instead are mechanically planted in a cozy-looking blanket. The soft, plushy fabric is made from recycled plastic bottles, and can be reused many times, a sustainability edge I appreciate enormously. To propel them upward, the seedling roots are misted with water and nutrients. This setup is registered as U.S. Patent No. 8782948B2. The first time I was shown the roots, I blinked in disbelief. They were so white, and pristine, so unlike any plant roots I had ever seen clinging to soil. AeroFarms' plastic plant beds are wide and long with rows of LED lights overhead that can be tuned to different hues along the color spectrum. Changing the hue while a plant is growing—red to blue, for example—can affect its flavor, color, and texture. There's little consensus on what style of indoor farming uses less water, but the aeroponic industry likes to quote that they use 40 percent less water than other forms of hydroponics. Everyone else likes to say that number is inflated.

Gotham Greens, an indoor farm that began on a roof in Greenpoint, Brooklyn, reports drastically lower water needs as well. CEO Viraj Puri said, "We use less than a gallon of water for every head of lettuce we produce." On a traditional farm, growing a head of lettuce takes more than fifteen times that amount. Gotham Greens' second greenhouse was built on the roof of a Whole Foods in Gowanus, Brooklyn. It's hard to argue with distribution that is literally feet away from the consumer. By the end of 2019, Gotham Greens was managing locations in five states, including Maryland, Rhode Island, and Illinois, totaling more than 500,000 square feet of greenhouses.

Greens from vertical farms sell for a higher price point than soil-grown greens, typically a dollar more per unit, which doesn't strike me as a path to getting people to eat more lettuce. If AeroFarms can nail the many layers of technology and scale up enough to lower its price per leaf, *and* build farms in areas that need it, *and* make the products available at a range of retail locations, then maybe AeroFarms, Gotham Greens, and other CEAs like them will reinvent agriculture. Instead of venture-funded greens for the wealthy, what if these startups had more than one revenue model baked into their business plan? There could be a tiered-price model, making greens affordable at all levels of income. In addition, the government could mandate that all health insurance plans offered under the Affordable Care Act be required to offer fresh food prescriptions*—directives written by doctors and paid for by health insurance. They already do this for other tech products like FitBit's; why should food tech be any different?

Founders universally talk to me of wanting to change people's mindsets toward fresh produce. One of the reasons we don't buy fruits and vegetables, they said, is because they spoil so quickly. A niche area in food tech that I've been watching with interest is how to protect fresh foods—keeping moisture in and oxygen out—that will extend shelf life. The best funded is Apeel Sciences, which you can find at Kroger markets coating its avocados, limes, and asparagus. The Santa Barbara startup produces an edible coating that is sprayed on fruits and vegetables to help them last for several days to a week longer than non-coated fresh foods. The coating is made from lipids and glycerolipids—things like fatty acids and glycerol—that exist in peels, seeds, and pulp of the foods we already eat. James Rogers, CEO of Apeel, speculates that once companies like his, and perhaps vertical farms, can extend their best-by date, it may break the hold packaged goods companies have on our shopping habits, and lessen the stream of snack foods fattening us up.

The Chef Allure

Sitting across from me during my taste test at AeroFarms was co-founder Marc Oshima. If ever I doubted whether his startup would succeed, it wouldn't be for lack of commitment or effort. Whenever I ran into Oshima at food conferences he was invariably typing on his laptop or juggling two phones. When he wasn't working, he was traveling for work.

As I nibbled my way through the just-picked greens, Oshima circled back to what the culinary world thought of his product. "The response we're getting from the tastemakers, the buyers, the chefs is like, 'I feel like my palate was woken up!' " This realm of food for the rich first reminded

* VeggieRx, a program to help patients with diet-related diseases, has operated in Chicago since 2019. Patients receive prescriptions from primary care physicians and dietitians that get them a weekly bag of fruits and veggies, and cooking lessons. The pandemic has led to a surge in demand.

me of how so many startups get their foot in the door: convince high-end chefs to use the product and then move down the chain to fast casual and grocery stores and, finally, if ever, to discount markets, bodegas, and dollar stores.

When I first visited AeroFarms in 2014, the company was growing greens in a single location: a former nightclub with black walls covered in glow-in-the-dark paint and what looked like giant white sleds planted with hundreds of seedlings. I chatted with the team and joined them for lunch, a big salad that was prepared in-house. From vertical farm to office desk? In November 2019, I trekked back to Newark. This time it was to see AeroFarms' commercial farm called "212 Rome," a seventy-thousand-square-foot former steel mill. When I toured through the building there was plastic sheeting up to protect the towering trays of produce from the space-expanding remodel. While big, this location will eventually be overshadowed by a *newer* farm yet to be built, a 150,000-square-foot building in Danville, Virginia, that the startup promises will "transform agriculture" and provide "safely grown produce that delivers peak flavor always." To run a farm of that size takes countless engineers, most likely remote, but only ninety-two people on site to plant, grow, and harvest.

To make the greens grow, AeroFarms engineers are tapping out computer code—called algorithms—to power grow cycles, LED lights, nutrient delivery, water, harvest date, and more. With the power of machine learning, these algorithms will, it is hoped, improve over time as more information is fed back into the data set. This includes on-site growers (humans) confirming sensor alerts—clogged hoses, nutrient levels, lighting malfunctions—a feedback loop that is necessary, but may one day spell the end of a human workforce. In a vertical farm, algorithms already largely run the show. And they are highly proprietary.

Before we made our way to the farm, I had to convince Oshima that I didn't want to see his special algorithms, an area of technology I could hardly steal or understand in a single glance. I assured Oshima that my interest wasn't in the underlying code, but rather the proposition of farms that lean heavily on technology. For example, does each and every green have its own algorithm? I asked. Or is there one for the greens that like it cooler, and another for greens that like it warmer? My mind swam with the possibilities. Oshima smiled, but didn't answer. Instead he told me that leafy greens have the potential to be an $8 billion industry. Then we wrapped up our tasting and drove two miles to 212 Rome Street.

On my way out, I was still hungry, so I helped myself to some free nuts.

The Pied Piper of Plants

"I like the products. I actually think they're tasty," Sam Mogannam texted me when I asked him what he thought of the greens he sold from Plenty. Mogannam owns Bi-Rite, a small chain of gourmet stores in San Francisco. Everything Bi-Rite sells is precious—almost too good to eat. "I'm not a huge fan of the amount of energy it takes to run the grow facility, and I have some concerns about the nutritional inputs used, but the water reduction, flavor, and quality are great. I do see a greater need for tech in areas where access to fertile land to grow greens is not available."

I live in California. Despite extreme temperatures, wildfires, and periods of drought, access to fertile land isn't our problem yet. According to the California Department of Food and Agriculture, the Golden State already supplies more than a third of the country's vegetables and two-thirds of our fruits and nuts. But that was actually *why* Plenty set up shop in the Bay Area; they wanted to compete head-to-head with the very best in produce. It didn't hurt that the founders were a short drive from Silicon Valley, and its hot pool of tech talent.

There isn't a single farmer on the staff of Plenty, but there are a few with "grower" in their title. Located in South San Francisco, just north of the airport, the workforce is three hundred strong, and growing fast. This includes a lot of engineers—some from Tesla—and an inordinate number of recruiters. It's not easy building a team that can grow a farm inside four walls. Plenty has raised almost twice as much money ($541 million) as AeroFarms, but in fewer rounds. Jeff Bezos is an investor. Plenty's humble offices don't telegraph big money until you see the yellow robotic arms, sit in on a meeting full of expensive engineers, or listen to a staff discussion on the minutiae of leaf width and shape, stem length, and something they called "tooth packing," which refers to how much food is left in your teeth. When I used the bathroom I noticed it was stocked with dental floss and mouthwash.

I wasn't shown around Taurus, the name of Plenty's first farm, which still requires humans, and still produces most of the greens sold to date. Instead, their then head of engineering walked me through Tigris, which is almost fully automated. After taking off my jewelry and zipping myself up inside a blue fabric onesie, I slipped booties over my sneakers and donned a hairnet. Stepping inside, we were at one corner of the massive big-box building, which still had room to spread out. We started at the beginning, where seeds are automatically planted in a potting medium, which isn't soil at all. Where a farmer might eagerly show off their healthy soil, at Plenty it's proprietary. Things it probably included: scentless shreds of coconut husk, perlite, and peat moss. Black potting trays made their way along a conveyor belt into an incubation room that is extra warm (the exact temperature of the room was a "trade secret"), and incredibly bright, to help speed the growth. It felt like Palm Springs in August.

Before I entered the incubation room, which was hidden under heavy black tarps, I was given giant black sunglasses to protect my eyes from the harsh white light. The lenses were so dark that it was difficult to see anything save for a dull pulsating glow. When I dipped the glasses down to sneak a peek, it seared my pupils like a solar eclipse.

After eight to fourteen days in this inner sanctum, the plants zipped out to a staging area where they were planted by a robotic arm into tall, skinny towers seven to thirteen feet tall that hold 40 to 150 plants. Another yellow robotic arm placed the finished towers onto an overhead zip line that moved the plants into their final stage, a grow room in which few people are allowed. Like a hospital room of newborns, there was a giant window that allowed visitors like me to peer in at the new life.

If you do an Internet search for "vertical farm," you'll see this iconic image. LED lights glow a pinky-purply hue, the towers and framework are pure white, and springing from tiny openings are rows and rows of flawless leafy greens. It's all completely different from the often highly romanticized image of the ideal farm, with chickens pecking and clucking, fluffy carrot tops sprouting from dark, loamy soil, and bumblebees buzzing around pollinating flowers.

Liz Carlisle, author of *Lentil Underground*, and professor in agroecology and sustainable food systems at UC Santa Barbara, has a lot to say about the benefits of traditional farming. "It's unimaginable to me that all of the benefits [of farming] could be replicated in a soil-less environment. We have yet to grasp the connection of microbial quality from the soil, and the plant to our [own] gut."

Another proponent of soil is Dr. Daphne Miller, a clinical professor at UC San Francisco with an interest in ecology. Miller told me that plants don't do as well in sterilized conditions when compared to a microbial rich soil. "We know it makes a difference to have an alive soil," she said. "The biggest argument for growing plants in soil is that we have tons of soil to grow food, it's just that we're abusing it, and growing the wrong things."

Long-term studies supporting these experts' views haven't been done, but there are countless doctors preaching for a diet of more organic produce, not less. Whether fruits and vegetables grown in a soil- and bug-free potting medium and given man-made fertilizers is the same as organic is up for debate. One pro-organic proponent is Dr. Michael Greger, who illustrates his stance using salicylic acid, the active ingredient in aspirin, one of the most common painkillers in the world, and also an anti-inflammatory phytonutrient. In plants, the compound is used as a defense hormone, and the concentration is increased by the presence of hungry bugs. "Pesticide laden plants aren't nibbled as much and, perhaps as a result, appear to produce less salicylic acid," he wrote. This micronutrient, which is found at low levels in fruits and vegetables, helps our bodies keep inflammation in check.

A *British Journal of Nutrition* study from 2014 looking at 343 peer-reviewed publications on organic food research found that organic plants produce more phenols and polyphenols that defend against pests, in turn nurturing higher concentrations of antioxidant compounds. As a result, they deemed that "organic fruits and vegetables contain about 20 to 40 percent more antioxidants than conventional produce." System health, an all-encompassing view where everything we do and eat is considered helpful to the bottom line, appears to be little considered by vertical farm founders who seem more focused on removing humans from the equation, and shortening the miles traveled by fresh foods.

The notion that we can replicate all that the history of traditional farming offers us inside a building is hard to fathom. We are becoming ever more distanced from where our food and water comes from, and the American investor culture, who sees every problem as capable of being solved with money, is helping widen this gap. The problems that regenerative agriculture solves—improved soil, planet health, animal welfare and nutrition—are also worthy of their attention. Maybe with their business smarts applied here, too, they can find the return on investment. Both solutions have merit and can be applied where they're best suited.

But back to the organic debate: Another vote for the higher levels of polyphenols found in organic foods, which have antioxidant benefits inside our bodies, comes from Dr. Emeran Mayer. Mayer, a leading researcher and clinician focused on the brain-gut microbiome interaction, and author of *The Gut-Immune Connection,* believes that organic farming in soil promotes a greater biodiversity in life—plants, pests, and microorganisms—that our own gut microbiome depends on. In his book, Mayer's thesis "is on the two reasons that plant-based food is healthier: one is fiber and one is polyphenols," he told me. "Vertical farm plants probably have the same amount of fiber, but I seriously doubt they have the same concentration of polyphenols." Not all plants are the same, and greens from vertical farms, which I've already noted are more water-filled and wispy, don't seem to pack in the fiber like the greens I get from the ground. What happens to our gut when we remove plants from soil?

Nate Storey, Plenty's co-founder and chief science officer, echoed the importance of farms in our conversations. "We're not competing with the field, we're filling the gap between supply and demand. The world has this way of turning our industry into an us versus them. Nothing could be further from the truth. Fields are tapped out. Let's build more fields without stressing the fields."

What's not tapped out are regenerative farms, where land is restored, nitrogen is replaced in the soil, and plant life (and bugs) is abundant and diverse. Some research shows that going this route—highly productive farms that have a light environmental impact—can maintain yield requirements needed for a growing world. Integral to this approach is that farms must plant legumes, which are a boon to crop rotation, and perennial crops, which lends well to multi-cropping—growing multiple crops on the same piece of land. By using these methods, we can reduce the yield gap between industrial agriculture and organic farming. If, for example, we shifted farm subsidies, and directed funds to these better forms of land management, the gap could shrink even further. Storey at Plenty could be right, too. Maybe his robot greens can complement existing and improved ecological systems?

Not all chefs are as erudite as Dan Barber. His book *The Third Plate: Field Notes on the Future of Food* (2014) is a guide for what our food system could be. In one word: flavor. We're not yet living in Barber's idealized vision, but the chef has it in his sights. He recently launched his own line of seeds, called Row7 Seed Company. Barber's brother, David, runs Almanac Insights, a food and tech venture fund that is an investor in Emergy Foods, the mycelium company from Chapter 2. Despite the connection, Barber the chef doesn't see technology as our savior. "There's a place for vertical farms in urban centers," he told me. And then in the same breath: "I'm not a pro-vertical-farm guy." While Barber sees vertical farms as the perfect city antidote, I see them in small towns with empty main streets and limited fresh food access.

"Where are the dollars going? They're going away from making healthy environments and nutrient-dense food and toward a reductive

A to B agricultural system. I wouldn't have a problem if it didn't purport to save the world from itself. We have a food system and agricultural economy that works in disastrous ways. But it doesn't have to, and there are ways to prove that."

In light of this conversation, I kept coming back to the vertical farm version of kale, possibly because it's so different from the one I know best, and eat almost daily: thick, toothsome, dark green lacinato kale. (I have to de-rib it!) My visit to Plenty started with a tasting of greens that could only be called delicate. I sat in a small conference room called Amaranth (named after a tiny, round, gluten-free grain) with growers, product managers, and quality assurance teams. We sat quietly munching away at trial batches of arugula. The team was hoping to determine if it was ready for the market. At the time, it was deemed too mild and watery. After several more grow cycles, it was ready. In August 2020, Plenty rolled out arugula and three other greens at about forty stores around the Bay Area.

While the team fashions its assembly line for plants—heavy equipment needs to be reengineered to work with leafy greens—it's racking up bills. I don't know what the burn rate (how much cash is required to keep the lights on) at Plenty is, but we may safely assume it's significant. In addition to the two farms that are side by side in South San Francisco, the startup is breaking ground on a third location in Compton, Los Angeles. At 94,875 square feet, the refurbished building can include more grow rooms to support a wider range of greens. Using sales data showing what people bought, the team at Plenty found that Los Angelenos bought salad mixes at a faster clip than anywhere else in the country. "They really like their leafy greens there," said Plenty CEO Matt Barnard.

After signing a nondisclosure agreement, I gained access to a one-hour meeting about Plenty's Los Angeles location. I later got the OK to write about it here. Seated around a long conference table were a dozen or so engineers. On the phone were many more. The topic was lighting design, and the group of mostly men discussed three design possibilities for the components, weighing in on the merits and downfalls of each one, including how much physical space the lights needed and whether they would be shipped fully finished on boats or be built on site.

How lights are used—what spectrum, responsiveness, shifts of color, feedback loops between sensors and plants—is illustrative of the data being parsed by high-tech farms. Being privy to the build-out of this new facility allowed me to see another reason that things at Plenty (and AeroFarms) move slowly. "We've built and torn down our farm over a dozen times in the last two years. Every day we get measurably better," Storey told me. To achieve their goals, the teams are constantly iterating—moving things one inch to the left and one inch to the right. The more decisions they nail, the more productive and profitable Plenty will be. (According to Plenty.) This iterative practice is mainstream when working with code and other virtual environments, but eventually these costs are passed along to people that can afford it.

Plenty has much to keep secret, but they allowed me more access than most. This included time with CEO Matt Barnard, who snacked on a plastic clamshell of Plenty greens while we spoke. In between my questions, Barnard rolled up fistfuls of undressed spring mix into tight cigar-shaped wads and popped them into his mouth. He extolled how much progress they had made. "We cut eighty percent of the total energy consumption of the farm," he said. This was by making better LEDs and by cutting about eighty-five percent of the people hours at the farm. These are "two of the most important metrics that drive the price of the food," he said. The third most costly: single-use plastic containers.

Despite having to nail all this minutiae around growing at high volume indoors, Barnard is certain Plenty will be profitable, with enough returns to please investors. I asked him how many years it would take for his investors to earn back their money. "In a number of years that investors find attractive," he said vaguely. After I pressed him again he narrowed it down to "a lot less than ten." Ten is traditionally the number investors look for to recoup their investment, but these startups are already on track to take more than ten. AeroFarms' ten-year anniversary was 2020, but the company says it won't be profitable until after they open their planned-for location in Virginia.

We would all be lucky if vertical farms could grow leafy greens with zero contamination. From 1973 to 2012, leafy greens were responsible for more than half of the fresh produce foodborne-illness-associated outbreaks in the United States. This is from a study by the University of Arkansas on pathogens in plant tissue grown indoors. I interviewed two of the researchers to learn more. Gina Misra, a molecular biologist who now works with Blue Marble Space on indoor farm education and outreach, told me how hard it was to gather information for her studies. "I did a survey of US microgreen [growers], and the reluctance of large suppliers to reply was difficult," she told me. "They're afraid of competition and they don't want to share anything. It's a little paranoid."

No one wants to admit to mistakes. At this point most product recalls we know about are from traditional farms, but the details aren't published about what happens to cause the outbreak. "They don't share data," Misra said. A lot of it is to shield businesses and their profits. "But how do we protect the people who provide food for us and keep them accountable at the same time?" asked Misra.

It was an area that I was given little access to. When I was reporting on a COVID-19 story for Bloomberg, I asked a vertical farm in Florida how they were purifying the air. They would only share that they used "a large air filtration system that scrubs the air for mold and fungus." When I asked how they were "scrubbing the air," I was told they weren't "comfortable providing that level of detail." An infectious disease expert told me that they were probably using vapor-based technology. He said I should talk to cannabis growers, who would be "more willing to talk." At Plenty, Storey said they used HEPA filters, which capture 99.5 percent of all particulate pollution, including viruses, bacteria, and mold. Whether it can capture COVID-19 is unknown.

For now, we can only hope that leafy greens grown indoors are safe. In light of the pandemic, I would think that two things may happen. Companies will have to share more about what they are doing to keep consumers safe, and, even though people getting COVID-19 from either grocery shopping or cooking at home has been largely unfound, food safety and handling could evolve from a federal regulation standpoint.

After safety, my big question was always: Why is the industry stagnating on lettuce? I eat salads daily, but does the world? Do they even want to? There are several reasons indoor farms start with lettuce. It's easy to grow and highly perishable. By shortening the supply chain—putting growing equipment on supermarket roofs or taking over empty buildings in urban centers—vertical farms can get us a better-tasting product. Most produce begins its journey in a refrigerated truck coming from California, but that means when they arrive in Ohio, for example, they're days or even a week old. Imagine if your greens only had to go downstairs or across the street?

If I believe what I'm told by these founders, they can control our food's growth with LED lights and targeted nutrients, even upping the polyphenols found in plant tissue—boons to our health. However, they're also reinventing form factor. Vertical farm plants are less fibrous and bitter. Many are sweeter. "If we want to change people's diets, if we want people to embrace consuming more fruits and vegetables, then it has to be easy to eat," said Storey. Plenty's end goal is snackability, but lettuce leaves replacing junk foods is a far-fetched Utopian vision. While I applaud the vision, it reads more like a health food writers' vision of a sci-fi world that will never exist.

Abandoning the Soil

"The evidence we have from the last century of industrial processing is that it's a problematic road to go down. If we only replicate the benefits that we can recognize, then we are still missing it," agroecologist Carlisle told me emphatically. "White bread is an example. Might vertical farms be the white bread of the twenty-first century? And we'll say, 'Oops, we left out something really important.'" Like the founders here, I want more people to eat their veggies, and any which way we get there is certainly a step in the right direction. But are we missing an opportunity to use massive amounts of funding to fix what's already (mostly) working—our farms? When the mission is to feed more people, why are methods of growing and varieties of crops kept as trade secrets?

Vertical farms have made growing crops a sexy new career path for engineers, but they've done little to address the fact that traditional farmers, the backbone of our food system, are aging out. According to 2017 census data from the USDA, the average age of farmers and ranchers is 57.5 years. If we still need traditional farms to grow most of our food, and our soil is depleted and our water supply sometimes tainted, then who is addressing the problems? Silicon Valley is easily wooed by robots and algorithms. When will they pay attention to people and soil?

At Cornell University, plant scientists teamed up with economists to assess the viability of indoor farming in 2019. Funded by the National Science Foundation, their report uncovered that traditional field production is by far the cheapest method of producing food. However, that cheapness doesn't include transportation costs, which are more expensive than the actual growing and harvesting. Barnard of Plenty has a saying: "There's no point in creating a more expensive version of the field." But that's exactly what they're doing at Plenty, AeroFarms, and other large-scale CEA farms like Bowery Farming and BrightFarms, which pulled in $100 million in new funding in October 2020, bringing it to more than $200 million in total funding.

Coming back to my main question of whether fresh foods grown indoors are as nutritious and healthy as those grown outdoors, both Plenty and AeroFarms told me their greens were the same or better in nutritive qualities. Neither shared any data supporting these claims. Regarding food safety, I wondered about pathogens inside indoor farms. Food safety experts consider vertical farms to be "built environments," which are equally as challenging to control as a farm, only with different variables. While vertical farms should have a cleaner water supply and no pesticides, there is always the potential for something bad to happen. Kristen Gibson, a professor of food safety and microbiology at the

University of Arkansas, said that in her circle they're thinking about how risks and human pathogens are going to change indoors. "It can go from the water being an issue, the people, the seeds that are used, to a variety of different things that aren't present in a conventional environment," she said. "In a built [environment] you are removing the pesticides, but there are still things that cause harm. You can't assume it's safer."

Misra, the molecular biologist, was more optimistic: "There's nothing that convinces me that indoor farming is less safe than outdoor farming, I think it's just less known," she said. On the other hand, she doesn't think that the industry will grow much bigger, because "people aren't chomping at the bit to get more greens." Despite that, both AeroFarms and Plenty are growing test crops of cherry tomatoes and strawberries, and by 2021 we could see them for sale. Strawberries are a high-value crop, so it makes sense that these indoor farms would make it their focus, although in California, where it's the state's sixth most important commodity, berries grow with ease—albeit propped up by the use of toxic chemicals. They also need to be pollinated, which means bumblebees in their precious indoor environments. Storey said he was already thinking of solutions for this natural process, and I imagined the costs involved in making a robotic strawberry pollinator; also how was it even possible to re-create something as vital to our food system as the bee, which is responsible in part for pollinating three out of four of our global food crops?

Solving the farming problem using technology fails to rescue our land and honor our farming legacy, and bypasses the people who have worked the fields for generations. Making kale as craveable as candy—or even the McDonald's-ization of kale—isn't priority number one when we look at our broken food system. I agreed with Carlisle when she rattled off three of her biggest concerns: Where are we getting our protein from (plant-based versus animal and industrial versus local)? How are we reducing our food waste (40 percent of the calories we grow are lost across the supply chain, including at our own homes)? And, how do we distribute food efficiently (we grow enough food, but it doesn't make it to everyone who needs it)? Answering these questions is more important than finding technology-intensive ways to grow *even more* food, especially when it's the same food grown for people who already buy it.

AeroFarms and Plenty have mostly proven they can make indoor farming work in dense urban cities in developed countries, but if we look at the operational needs of CEAs—farms that depend on engineers, a 24/7 electrical system, a cold-chain distribution network (trucks to deliver goods, which keep perishable foods fresh), and a steady stream of plastic packaging—they look less and less sustainable. They certainly don't look like something that could work in sub-Saharan Africa or in much of India.

Even so, the lazy eater in me, the one that wants convenient and delicious—even when it's packaged in plastic—is occasionally persuaded. Plenty's tomatoes, Storey assured me, would blow my mind. "I'm a total tomato snob. We've been breeding tomatoes for a little over a year, and last week I had two of the best tomatoes of my life." They sounded almost mythically good. But I wondered how much they would cost? And the bee problem that I had imagined teeny-tiny robotic fingers for, was solved when the team settled on using a European bumblebee that doesn't produce honey. Strawberries are probably what seduced new investors. "There are strawberries out there that taste as good as the strawberries that we can produce indoors. The difference is that we can do it every single time, all year long," said Storey. This is the hope or promise of all technological innovation: cost-effective, consistent results, at scale.

Chapter 8

Cell-Based Meat

Will Animal Analogues Make It Beyond an Elite Niche?

Steak in Space

In the early days of space travel, astronauts ate things that resembled baby food. Imagine three meals a day of puréed carrots? Even more distressing was a dish they ate in space called "Appetizing Appetizer." It came in a can and looked and tasted like cat food. Another option, the food pellet, seemed to be the easiest way to feed people in orbit, until they arrived back on Earth malnourished and hungry, convincing NASA's food scientists that it did matter, and it did offer significant psychological rewards. In its history, the space agency owned up to its misguided attempts: "The food that NASA's early astronauts had to eat in space is a testament to their fortitude."

In its ongoing quest to find solutions for how to sustain astronauts in space, NASA funded a research project in 2002 aimed at growing edible muscle cells "in-vitro," a term used in the lab-based movement that means producing cells outside of their normal biological home. First up were goldfish cells grown outside of a living fish. Next were turkey cells. Eventually the projects were scrapped due to the cost of scaling, and perhaps a small amount of ick factor. However, the goldfish project inspired the beginnings of New Harvest, a nonprofit dedicated to promoting cellular agriculture. There are dozens of startups vying to be the first to bring cultured meat to market. A sampling of their names: Future Meat Technologies, New Age Meats, SuperMeat, Integriculture, Higher Steaks, Aleph Farms, and Meatable. Few stray far from the meat, or the farm.

How they're doing this work is not too dissimilar whether the startup is in Israel, California, or the Netherlands. Cultured meat is made of animal cells that are grown inside a lab. These cells are fed nutrients to help them proliferate. When there's a big enough mass of cells, they're formed into what is hopefully a close match to the real thing: chicken, beef, or duck. Founders present with one overarching commitment: to do away with the detrimental environmental practices that come from raising animals on an industrial scale, and our reliance on eating them for dinner. Many of these companies have impressive amounts of funding, but there are hurdles at every stage making this promise still years, if not decades, away from full-scale, feeding the billions reality.

Despite this, one company has received approval to begin selling cultured meat. Just CEO Josh Tetrick said his company would have cultured animal meat to market in 2019. It wasn't quite then, but in December 2020, Just announced that it had received regulatory approval from the Singapore government to begin selling a hybrid chicken nugget made of both cultured chicken cells and plant-based protein. In a four-day trial at 1880, a private club in Singapore, the nuggets were plated up for invited guests at a cost of $24 per person.

Animal tissue has been grown in labs for decades, but it only made the leap to food in 2013, when Mosa Meat, the Dutch startup founded by Mark Post, produced the world's first cultured hamburger. That burger cost $330,000 to make. At the time, the company said that "one sample from a cow can produce enough to make eighty thousand quarter pounders." Most founders are confident about the future of cell-based meat. Post is the man widely considered to have launched the cell-meat movement, and has been working on growing meat in labs for fifteen years. Whether from Dutch prudence or a long career of attempts, Post is far less guarded than most founders about the uncertain road ahead. In an issue of the journal Food Phreaking called "What Is in In-Vitro Meat," published by the Center for Genomic Gastronomy, Post wrote that producing consumable meat in the lab was a potentially, yet-to-be-proven, resource-efficient possibility. In 2020, Mosa raised $75 million, but it's clear there are no guarantees coming from Post.

How Does It Taste?

On the other side of the world is Memphis Meats, which is so confident in its technology that it's already mocked up packaging. The box I held in my hand looked legit, like anything I might pick up in the supermarket. Attractive branding—a close-up image of a stoneware plate with a grilled piece of chicken breast atop a bed of lacinato kale and a few curls of purple onion. A little window allowed me to peek inside—at a skinless chicken breast covered in plastic. "Made with love in California" was stamped on the box. On the back, under the nutrition facts panel, were the ingredients, so common you might ignore them—sea salt, chipotle pepper, sugar, and garlic. Nothing unusual except the first one: chicken (cell-based). And it really was a chicken breast, but not a breast butchered from a dead animal; it was grown from cells in a lab by Memphis Meats of Berkeley, California. The actual chicken who contributed the cells may still exist on a farm somewhere.

The second floor of Memphis Meats' headquarters opens into a giant kitchen worthy of a cooking show. Behind the range stood food scientist Morgan Rease with a hip Brooklyn beard and wearing an apron. The air was fragrant with sautéed mushrooms. My nose twitched and I automatically began salivating, even though I had recently eaten lunch. "Is there anything you don't eat?" Rease asked. My list of dislikes is short, but while writing this book my motto was: "I'll eat anything."

While I took in the chic cabinets and massive kitchen island, Rease and I waited for Uma Valeti, the CEO of Memphis Meats. Before joining the food tech revolution, Valeti was a cardiologist, a lifesaving career, but with his new endeavor he hoped to save even more human lives and stop animal cruelty. As a child growing up in India, he recalled attending birthday parties alongside seeing kitchen staff gutting animals to supply the "joyful" feasts. After medical school in the United States, Valeti stayed put. In addition to clinical work, he had a research lab at the University of Minnesota where his patients had one thing in common: massive heart attacks. Stem cells were one of the treatments, and the founder began to wonder if he could make humans healthier—what about the food we eat? This idea percolated until he was introduced to his now co-founder Nicholas Genovese, who has a PhD in oncology. It was the catalyst the pair needed to jettison their medical careers for the high stakes and highly uncertain future of cultured meat.

Five years after leaving medicine, Valeti still has that doctor vibe—measured, well-spoken, and with an assurance that made him the ideal person to go knocking on doors asking for large amounts of money. Every cell-based meat company cringes at hearing their work called fake, but in the early days it was still outlandish enough that most investors showed Valeti the door. Nonetheless, in its earliest fundraising, called a seed round, Valeti pulled in just over $3 million. "This industry had never been funded," he said, until he proved to investors that he could grow animal cells rapidly. Today the vision has grown into a company of more than sixty staffers—from animal activists to environmental champions and even meat eaters—racing to become the first company to commercialize cell-based meat tissue.

On our way to a conference room, Valeti paused in front of a company timeline painted on a wall next to the bathroom. Many dates pegged to Memphis Meats are landmarks, such as the founding of the company (Valeti considers his to be the first cell-meat company in 2015), the first meatball (produced in 2016 at a cost of $1,000), and its series-A round in 2017 for $17 million, the largest amount of funding in cellular agriculture at that time.

While most cell-based meat startups focus on a single species, Memphis Meats is agnostic, and maintains that its platform is capable of growing all types of cells and tissue. Scientists in its seventeen-thousand-square-foot headquarters have grown beef, chicken (the most widely consumed meat in the United States), and duck (the most consumed in China) and served them to more than one thousand people.

Valeti walked me back to the kitchen, where Rease was pulling a small piece of chicken off his sauté pan. He placed it on a cutting board, and gently sliced against the grain. Valeti urged me to watch: "The cutting and the texture is something you should notice as Morgan cuts [the chicken]. This cuts exactly like a piece of chicken."

On a plate next to Rease were two large gold spoons layered with sauces. "No one eats plain chicken," said Valeti. Tell that to the bodybuilders, I thought, who Valeti would certainly want enjoying his products. One spoon held the flavorings to make it a single bite of chicken piccata. The other held chicken satay with peanut sauce and house-made gingered pickles. Next to the two spoons there was an undressed bite to taste plain. I looked down at the plate, and then over to the chef, and across to Valeti. David Kay, Memphis employee number one and the head of communications, was nearby taking photos. Eating samples in front of people who have tirelessly worked to reinvent our food supply was one of my more uncomfortable professional experiences.

I sliced. They watched.

Valeti was right, it did cut like chicken. I put a half-inch morsel in my mouth. Like traditional chicken, it had tug and chew, something for my teeth to grab on to. I could feel the strands of muscle in my mouth. It was also very dry, missing the juicy moisture I wanted from chicken. Valeti assured me there were fat cells in addition to muscle cells, but I couldn't discern them. The meat itself had flavor, but the oil it was fried in made a bigger impact on my senses. Later I was told that the meat I was eating had been grown from cells taken from a chicken egg. A familiar origin story—chickens come from eggs—but I wondered how many people would be willing to switch?

Next, I put the chicken piccata bite in my mouth. For those who crave meat, it was delicious, and far tastier than the plain bite. The texture of the chicken played well with the butter, lemon, and capers. Thinking back to my sensory evaluation lessons at MycoTechnology in Denver, I slid the food around in my mouth, giving my taste buds time to absorb, and my mind time to think.

As if I were on an episode of *Top Chef*, the men watched intently for my reaction. I avoided their gaze, wishing I could jot down some notes. I said "Wow" a lot, which gave me time to think. "It tastes healthy," I said. It probably wasn't what they wanted to hear, but it was honest. Most important, they had nailed the texture—the sine qua non of all meat analogues. "The texture is amazing. Impressive," I repeated.

With only a few small tidbits to consume, I found it hard to visualize a whole chicken breast on a plate. Valeti assured me, however, that they had made "full format chicken," and hosted frequent tastings—even a recent event for a hundred people. Chefs had tried it, he said, and told him: "I could put this on a plate right now and it will be the most tender, flavorful thing on our menu." I tried to imagine what Thomas Keller or Alice Waters might do with it. At Chez Panisse, the apex of California-local cuisine, Waters might baste the chicken in a sauce of morel mushrooms and serve it alongside roasted red thumb potatoes and sautéed chard, the bites of cultured chicken becoming a sideshow to the far prettier plants. Maybe Chef Keller would sous vide his chicken in a bordelaise sauce made with dry red wine, bone marrow, butter, and shallots. He'd set the chicken on a plate atop wilted arrowleaf spinach and Nantes carrots.

I snapped back to reality. Whether cultured meat could be seen as a gustatory delight—say Thanksgiving dinner—seemed far off. Waters would never accept meat that came from anyone other than from a farmer she knew. And even the experimentally minded Thomas Keller would be perplexed at how to feature it on the menu at The French Laundry, his Michelin-starred restaurant in Yountville, California. "Cultured meat from Memphis Meats," it could read. Maybe the word "cultured" would help them sell it for more? Servers would have to instruct diners that the company was from Berkeley, not from Memphis, and employed scientists instead of butchers.

But instead of unattainable dining experiences, what if cultured meat startups acted like their goal was feeding everyone? They could find more approachable culinary experiences like the buttermilk fried chicken at chef Tanya Holland's Brown Sugar Kitchen in West Oakland, California, or the BBQ brisket bowl at chef JJ Johnson's Field Trip in Harlem, New York?

Chef adoption is mission critical in the future food movement. By serving Memphis Meat products in their restaurants—which hasn't happened yet—chefs lend instant credibility to these unvalidated foods. "We feel like it's ready to go to market," Valeti told me. "But we always see room for improvement." With that, Valeti picked up his phone, plugged in his headset, and jumped on a call. Kay escorted me to my next interview.

How Do They Make It?

Making a burger without a cow is a daunting process. Here goes: Live cells—or cell lines—taken directly from an animal will only grow a limited number of times. They can be turned into the ideal version,* an "immortal" cell line, only through complex steps in the lab. This version of almost animal-free cell lines is what every cultured-meat startup is working toward, and one of the big hurdles. True story: The cell lines that Artemys Foods uses are from a bull named Future, who has a nice life on a farm in Ohio. The startup told me that genetically the bull is "amazing," with cells that are "robust."

To get cells from Future the bull, for example, the first step is to perform a "punch biopsy." A small tube of cells is removed from Future using a cutting tool that looks like a cross between a pen and a needle for blood. This is how human biopsies are done. It doesn't hurt that much, but it's not fun. Unlike you and me, the animal can't consent to the procedure.

* Stem cells are considered immortal, which is why they are such a large part of medical research.

One hurdle to growing cultured meat is finding the right starter cells. The best ones are stem cells from muscles. Inside an animal they foster new muscle growth; outside the animal they can be "programmed" to grow. To be productive and successful for food development, cells must be self-renewing, ideally in perpetuity. Once identified, these starter cells, or cell lines, can be labeled, packed up in vials, and stored in liquid nitrogen 275 degrees below zero Fahrenheit. To propagate, cells are placed in a medium containing nutrients and growth factors. Growth factors are found in serum, which is the clear liquid in blood. Early on, companies used fetal bovine serum, or FBS.

Biomedical research depends on FBS, which comes from the blood of a cow's fetus. FBS is expensive, which the pharmaceutical companies can afford, but it's not made in quantities to support mass scale production. FBS contains albumin—a simple protein—plus a small amount of amino acids, sugars, lipids, and hormones. Because it comes from cows, it's frowned upon—remember that almost every founder in this book is vegan. But there are other reasons for the frown: The supply chain for food production needs to be sourced from simpler, cheaper ingredients. There are no alternatives, yet, that are as effective as serum in promoting cell growth. (Cultured meat startups are working on them, but little has been made public so far, and few outside of academia are collaborating and sharing information.) If the sector is lucky, one of New Harvest's many funded research projects focused on animal-free mediums will get there first. The nonprofit is based on the idea of shared knowledge.

To put a piece of meat on a plate, companies need to grow trillions upon trillions* of cells inside steel tanks called bioreactors that range in size from 1 liter to 100,000 liters. (Memphis Meats uses what it calls "cultivators." Based on internal designs, these tanks will be custom built.) If everything goes well, the cells—myotubes—will differentiate into muscle cells. Depending on the meat being produced, say steak versus ground

* A 3.5-ounce piece of traditional meat is made up of about ten billion cells. To make the same sample of cell-based meat would take roughly ten trillion cells. (But who's counting?).

beef, an underlying scaffold is needed for tissues to adhere to. You can even think of your own skeletal body as a scaffold for cells. Amy Rowat, a professor of bioengineering at UCLA, received a grant from the Good Food Institute (GFI) to work on micro-scaffolds that might create the marbling in cultured beef. "Marbling" is a fancy way of saying "streaks of fat."

"Fat is important for texture and flavor. Our goal is to develop a cultured meat scaffold that can promote marbled meat, where fat is interwoven within the muscle," said Rowat about the work in her lab. The scaffolds her lab is creating will have different regions of stiffness. Fat needs softer cells, and muscle cells need stiffer fibers.

Rowat's lab also has a grant from New Harvest, which has funded other researchers working on the structure problem. One scientist is even working through this problem using spinach leaves. Imagine a meat eater being told their synthetic steak had the veggies built in?

I've wondered a lot about what the cells are fed on. When I suggested that consumers would want to know how their meat was grown, the response I got was either "We can't share those details," or "No one knows how the animals they eat are fed. How is this any different?" While I disagree, I know they exaggerate! When I asked again, I got this: "It's identical to the needs of live animals."

Animals need what humans need: essential amino acids, fatty acids, carbohydrates, vitamins, minerals, and water. The king here is a good source of carbohydrates—a required fuel source for virtually any living thing. This can come from grains, the starch in vegetables, or from something as simple as white sugar. (Whole foods offer the widest range of nutrients.) After landing on the "food" supply, startups must determine how to supply chemical signals, which are required to stimulate cellular growth. This relies on hormones. Making hormones—like insulin—without an animal (or a human) is expensive and usually requires genetic engineering. This is a major roadblock for cell-meat startups.

Meat cells are alive. They don't die until they're harvested—another tricky piece of the puzzle. Technicians supervise the temperature, water activity, oxygen, nutrients, and pH value (level of acidity or alkalinity).

The entire process takes about a month. When the "meat" is harvested or "killed," its growth is stopped. Finally, these real meat cells can be formed into a tasty piece of something. This step includes the basics of food processing, things we might consider easy: seasoning, shaping, and cooking.

Next on my tour was Eric Schulze, vice president of product and regulation at Memphis Meats. When I sat down to interview him, Schulze confessed that he smokes his own traditional meats and barbecues every weekend. "I don't want to give up eating meat," he told me. "I am acutely aware of what I'm doing and I want to unburden myself of my own guilt. That motivates me." A towering redhead, Schulze emits proclamations like gamma rays. When I mistakenly referred to what they were making as protein, he corrected me: "Meat is much more than protein. We're creating a tissue. Meat. Not a protein."

Schulze was previously a regulator with the FDA, and speaks fondly of his time at the federal agency. I asked him what consumers should want to know about the safety of cell-based meat. "Let's say a cereal company started making a better cornflake, using a faster machine," he said. "They don't have to put on the label 'Made with a ten-times faster cornflake-making machine.' That's not illegal because the product is fundamentally the same product nutrition-wise. The way it performs, its identity, the way it looks," he said. "That's the same thing here."

He was seriously oversimplifying, so I nudged Schulze for more. "Every producer of food in this country can voluntarily disclose what they would like their consumer to know about their product if they think it's true." He said that telling consumers that meat from Memphis Meats is cell-based has value, because they can point to things that are good for humankind like the sterile environment it's grown in, and the lack of antibiotics.* He also said that while the company was still undecided

* In September 2020, at a virtual Future Food-Tech conference, Uma Valeti said that "cell-based has a role in reducing future pandemics." As for meat made without antibiotics, he said Memphis Meats "expects not to use antibiotics," but with untested technology that has yet to scale up, it's not a guarantee.

about the level of transparency, they are "open to voluntarily disclosing the production method."

Experts in the industry are hard-lined about this. In a talk about the road to consumer acceptance for cell-based meat, Greg Jaffe, director of biotechnology at the Center for Science in the Public Interest (CSPI), told the conference attendees, "We have the framework but not the details." Not only do we not have the details, but there's a long list of possibilities for how to produce meat replicas that include gene modification, genetic engineering, cloning, and fermentation. Any products, Jaffe told the virtual audience from his home office in McLean, Virginia, at the Industrializing Cell-Based Meats conference, would need to be independently verified, and product labeling needed to be "truthful, neutral, and informative."

In food, "truthful" can be bent at will. Like declaring "fat free" on a box of licorice that's only made of sugar. "Informative" seemed another word that could be bent to a company's will. I found transparency sorely lacking in many of my interviews for this book, and it's hard not to let that translate to an increase in my own suspicions.

What's the Motivation?

When I talked to founders in Silicon Valley, money was—believe it or not—usually the last thing they wanted to discuss. Ben Wurgaft, the author of *Meat Planet* (2019), felt similarly. "I kept wanting [to interview] people who would admit to wanting to get rich," he told me. "People want to unite a moral imperative with a market imperative. The primary motive for speed could be animal suffering, but it could also be money," he said.

Like me, he felt many of the founders were sincere. They pointed out the moral dilemmas of animal welfare. Meat consumption in developed countries is steadily rising, and we'll need to slaughter ever-increasing numbers of animals to support this habit. Numbers don't matter from the moral point of view; it could be one or one million that get eaten. In fact, thirty-six billion animals were killed for food around the world in 2020. Valeti said he stopped eating meat because he wanted a method that he could get behind. Our current way isn't it. In light of COVID-19, which engulfed the world in a long year-plus of health concerns, and which was linked by experts to our consumption of animals, anything that reduces our continued encroachment upon wildlife habitat is worth consideration.

Next on founders' checklists was how unsustainable animal agriculture is, how bad it is for the environment, and the poor conversion rate of feeding animals to feed humans. The ROI—return on investment—of feeding crops to animals in order to grow protein to feed us is "highly inefficient technology." For their finale, they offered up a much-repeated statistic, couched in the question: "How will we feed a world population of nine billion by 2050?" With limited arable land, younger generations uninterested in farming, and a growing need for protein, these founders considered cell-based meat to be the top priority in saving the planet and humanity.

Transforming to healthier diets will take effort. In a report by the EAT-Lancet commission, a group of international scientists recommended that we double our consumption of fruits, vegetables, nuts, and legumes. Perhaps more challenging was that they said we also needed to reduce consumption of red meat and sugar by more than 50 percent. Doing this, they said, would "confer both improved health and environmental benefits." Dr. Brent Loken, a scientist on the commission, said that without these changes, food emissions were expected to double by 2050. "Food-related emissions would likely result in the world exceeding the global warming limit of 1.5 Celsius in 30 to 40 years," he said.

This opinion is widely held. "The idea that we can eat for our own health while ignoring the health of the planet is, quite simply, extinct," said Dr. David Katz, the founder of Yale University's Yale-Griffin Prevention Research Center, an author on nutrition, and the founder of Diet ID, a mobile diet-assessment and behavior-change tool.

In 2019, when my Instagram feed was taken over by images of the Amazon burning, I knew it was fuel for the arguments of all the founders I spoke with. The *New York Times* called it "ecological arson." Cattle ranchers were burning down the largest, most biodiverse rain forest in the world. There was a huge public outcry. If cell-based meat had already been available for purchase, ads for it would certainly have proliferated in my Instagram feed alongside images of the devastation.

But while cell-based meat is still at a startup stage, plant-based meat is already here. You'd be hard-pressed to find someone who hasn't eaten a Beyond Meat or Impossible Foods burger and declared it delicious. Other companies are launching their own versions. With rapid advancement in plant-based versions of burgers, bacon, and pork, are these more complex cell-based foods even necessary?

One of Memphis Meats' early investors is Seth Bannon of Fifty Years, an early-stage venture capital firm. A longtime vegan, Bannon has watched the evolution of the industry since 2014, and puts Valeti in the early founder category, the "die-hard believers that the existing food system is broken." With money invested in both plant-based and cell-based startups, Bannon wants both areas to succeed. "We are super-bullish on plant-based," he said, noting that there was room for one hundred times growth in the category. But he also thinks that cell-based meat will do really well. "There's room for dozens of ten-billion-dollar companies in each market," he said.

Where I see plant-based as the simpler solution, as long as we can make it healthier, Bannon finds them both gaining traction. "In one way, plant-based is ahead. They're in market, they have partnerships and happy consumers." But, ever the encouraging dad who wants to see all of his kids succeed, he also felt that in another sense cell-based was ahead. "If you give chefs Impossible or Beyond, and Memphis, they'll say that Memphis is closer in terms of replicating what people want." The problem? It's too expensive. While some of us may be able to order duck kabobs at a popular San Francisco restaurant in the next year or two, a commercial package won't be available in the meat aisle for years to come.

In Memphis Meats' early days, potential investors looked over the business plan and declared it science fiction. Bannon thought otherwise, but few others were ready to invest. Then, when plant-based competitor Beyond Meat was listed on the New York Stock Exchange on May 2, 2019, the public markets weighed in so enthusiastically* that private investors finally decided cell-based meat could also make them money. In January 2020, Memphis Meats closed a series B for $161 million and the company received so many term sheets—preliminary financial agreements between startups and investors—that it cherry-picked its favorites and showed the rest to the door.

Plant-based and cell-based may seem at odds with each other. But not only are they tackling the same issues—the damage of animal agriculture on the environment—chances are highly likely that cell-based startups will adopt a hybrid approach and create foods made from both cell-based tissue and plant protein. In fact, one new startup is doing just that. Artemys Foods, located in San Leandro, California, plans to create blended burgers because they're the most widely eaten in the United States. The founders of Artemys told me that even adding only 10 percent of cultured cells to a regular plant burger "enhances the flavor dramatically." Like other synthetic biology believers, Artemys believes that its halfway approach will get to saving the world faster. March said, "The only way to change how the majority of people are eating meat is to be able to give them real meat."

The dual approach is even being adopted by the traditional meat industry in the form of chicken nuggets blended with veggies, and beef burgers cut with mushrooms and rice. Their reasons may be different—to lessen their impact, to help cut costs, and to capitalize on the plant-based trend—but the framework is the same.

Bruce Friedrich wants to see both options win, too. The executive director of the GFI, a nonprofit founded in 2015, is a force in the

* In its first public offering, Beyond Meat's stock was priced at $25 a share on Nasdaq. Shortly after noon, it traded at $46, and at market close was $65.75, for a gain of 163 percent. It was the biggest IPO for a US company valued at more than $200 million since 2000.

movement gaining momentum in steering our diets away from animals. True story: As part of an animal-rights publicity stunt, Friedrich streaked at Buckingham Palace right before an appearance by George W. Bush in 2001. After decades spent trying to convince people to save animals, he's now hoping to distract us with foods that are as good as or better than traditional meat. "There is absolutely no way meat consumption will go down. So we need to make meat better. We need to make it from plants. We need to cultivate it," he said.

Funded initially by Mercy For Animals and then later supported by people like Suzy and Jack Welch—longtime CEO of General Electric—and Dustin Moskovitz, one of the lesser known co-founders of Facebook, GFI supports research in both cell-based and plant-based alternatives to animal products. The nonprofit funds research and lobbies in Washington, D.C., and works with regulators to pave the way for change. Friedrich told me his "pie in the sky" wish was to see billions of dollars in government funding for open-source R & D into meat alternatives. He'd even be happy if it was the Chinese government that made it happen. "They would have bragging rights for the rest of time if they were the government that eliminated industrial animal meat from the face of the Earth," he told me. The longtime vegan also started New Crop Capital, a for-profit private venture fund that invests in dozens of these companies—both Mark Post and Uma Valeti are advisors to GFI. While billions from governments seem a ways off, investor money is being funneled into these burgeoning food sectors, and Friedrich is responsible for much of its traction.

Is It Good for the Climate?

Mass production of cultured meat relies heavily on water, energy, and crops. Knowing just how much is unlikely without details from the companies. Will creating a pound of cultured beef use fewer natural resources than a pound of ground beef? In a 2020 article titled "Bridging the Gap Between the Science of Cultured Meat and Public Perceptions," from

UCLA researchers Tomiyama, Rowat, et al., one biopsy from one cow could theoretically provide one billion beef burgers in one and a half months. To make the equivalent number of burgers with conventional farming, we'd need eighteen months and half a million cows. More than 90 percent of the world's population eats meat. Will these cell-meat companies produce foods that can make everyone happy?

Cultured meat is being tackled all over the globe, and not just in the lab. Located in Rehovot, Israel, and founded in 2017, is Aleph Farms. In addition to lab work, the Israeli startup has launched its cells into space. In a collaboration with a Russian 3-D printing company, on September 26, 2019, the cells launched into space nestled safely in the arms of Russian cosmonaut Oleg Skripochka. His target was the International Space Station (ISS), where he would try to prove that cultured-meat cells could successfully grow "in the harshest conditions possible—without land or water resources," said Didier Toubia, co-founder of Aleph Farms. Toubia's other questions: Would the cells survive the launch? Would the temperature be OK? Would they grow in microgravity? Next, could the cosmonaut assemble the tissues under microgravity in the short time aboard the ISS?

And, ta-da, it worked. The cells interacted easily with each other. Once back, the ISS experiment informed Aleph's decisions on what technologies to use in its pilot facility. Its foray into space helped the team inch closer to its goal of developing a circular (no waste), closed loop production method on Earth. That production method will help Aleph Farms bring their thin steaks to market by the end of 2022. Or early 2023. For now, it's anyone's guess.

But back to my earlier question on whether we should focus on plant-based meat analogues over cell-based meat. According to a 2018 study on reducing food's environmental impacts by Joseph Poore, a researcher at the University of Oxford, England, "A change in global dietary habits from meat to plants would be enough to offset the expected increase in world population." Looking through a different lens, that of life-cycle carbon analysis, Asaf Tzachor at the University of Cambridge noted that if everyone switched to a plant-based diet we would see a 49 percent reduction in greenhouse gas emissions (GHGs).

Although factory farming is a leading contributor to GHGs in the United States, the biggest producer is transportation, listed in an EPA report as causing 28.7 percent of total GHGs in 2019. Next was energy (27.5 percent), industry (22.4 percent), and finally agriculture (9 percent). While this number is up from previous years, it's down 2 percent from a decade ago. (Globally the number is 14.5 percent.) This is not something to cheer. We need ever more sustainable systems.

While traditional agriculture results in all three GHGs—carbon dioxide, methane, and nitrous oxide—cultured meat emissions are almost entirely CO_2, which comes from the energy needed to power the plant. On the surface, that little nugget makes us think that cultured meat is far superior. In a study on the climate impacts of cultured meat and beef cattle by University of Oxford researchers John Lynch and Raymond Pierrehumbert, the pair found that "GHG emissions per unit of cultured meat are uniformly superior to that of beef." But not so fast.

Initially, the researchers wrote that "cultured meat results in less warming than cattle," but "the gap narrows in the long term and in some cases cattle production causes *far less warming*, as [methane] emissions do not accumulate, unlike [carbon dioxide]"—the primary gas from cultured meat. The best path for cultured meat to be better for the environment than traditional meat is to ensure the systems creating it are largely, if not completely, reliant on renewable energy. Memphis Meats has signed a lease on a production facility near its Berkeley headquarters, and the staff are working to design and build the systems that will churn out our new food. No word whether it will it be run on renewable energy.

Researchers Lynch and Pierrehumbert also cautioned that the trends are far less clear for other animal products, chickens especially, which are environmentally more efficient converters of crops than cows. This means that when we feed crops to chickens, they convert that feed to body weight, which equals calories when we eat them. Chickens need

1.7 pounds of feed to produce one ounce of meat versus a cow, which needs 6.8 pounds of feed. If we wanted to pick one of these proteins to be the true climate champion here, it would certainly be a plant-based analogue of chicken, which is incredibly resource efficient—simple plants converted into "meat." While we have faux chicken nuggets in spades, a breast, thigh, or leg has yet to be perfected for the marketplace.

The rhetoric put forth for marketing to venture capitalists, angel investors, and consumers is that cell-based meat is going to save the world. Christy Spackman, assistant professor in the School for the Future of Innovation in Society at Arizona State University, pointed out the errors in thinking this way. "I would argue that this isn't the lesson history would teach us. Industrialization made it harder for us," she said. That industrialization, which we initially cheered along, caused the problems we have now—confined animal operations that result in toxic runoff that can't be reincorporated into the land, for example. Spackman felt it was urgent that we consider the systems that supply the things that are needed for cell-based meat before new foods are produced at an industrial scale. Keep in mind that industrial scale is what every cell-based startup aims to achieve right out of the gate.

For his part, *Meat Planet*'s Wurgaft thinks it's a good thing to see our appetite for industrial meat shrink, but he hopes that comes by way of a public that is allowed to make an informed choice rather than through government intervention or market pressures. "I think it's important we don't abandon hope that people can, in fact, take responsibility for their choices."

Wurgaft and I first met on Twitter. In 2017, we met in person at the New Harvest Conference held at MIT, where he was a visiting scholar at the time. What I liked most about the wild-haired, bespectacled writer was that he wasn't a food guy, he didn't work for a startup, and he wasn't an investor. He ate meat. He had zero skin in the game. Part history nerd and part philosopher, Wurgaft comes at the topic of cultured meat without the dogmatic belief that eating animals is bad, or that eating vegan is the only way forward.

Safe Enough?

Cultured meat is a world away from small-batch gourmet foods made in someone's home kitchen and sold at the farmers' market. Cultured meat doesn't have that luxury. Can you imagine seeing a piece of chicken on a folding table at the market with a price tag of $1,000? Instead, the companies are going straight from the lab to mass production, working to drive the price down low enough to attract mass consumption. It's not good enough to sell to the rich, which is where it will sit for a while until they get costs "low enough," which means not just affordable, but cheaper than cheap beef. To get there, they need to develop a scalable process for a manufacturing system that's never been tried before, and significantly lower the cost of a nutrient that has never been created before.

While we're years from seeing cell-based meat widely available in supermarkets, a small group of US companies have formed an association to support the growing industry. Called AMPS, or Alliance for Meat, Poultry and Seafood Innovation, it includes Memphis Meats, Just, Fork & Goode, Finless Foods, BlueNalu, Artemys Foods, and New Age Meats. Because the group is focused on American regulatory pathways, only US companies can join for now. That is if they can pay the $50,000 membership fee.

Regulatory hurdles are often listed as one of the top barriers to market. (The others are consumer acceptance, growth medium, and legal battles from existing lobbying groups.) Almost universally, these food-tech companies began working with the FDA before they had a product to share, which is something the regulators want; and these meetings are all behind closed doors. Regulatory approval is crucial—a high-stakes effort that will include multiple levels of oversight, inspections, naming, labeling, and more. Complicating matters, the USDA oversees meat, including beef, chicken, lamb, and pork, and labeling, while the FDA oversees everything else, including fish (but not catfish!). This means cell-meat companies need to work across two agencies that aren't known for working well together. While not fully ironed out, so far the two agencies plan to share oversight responsibility. Valeti remarked,

"They recognize that this is one of the biggest technical opportunities in food." The FDA will regulate the science (and fish), and the USDA will regulate inspections of the actual products (except fish) before they hit the market. The USDA has a long history of daily food inspections. The FDA does not. Once production ramps up with giant bioreactors producing trillions upon trillions of cells, the FDA may only occasionally drop by to look the place over.

Many insiders say that once the United States approves cell-based meat, the rest of the world will follow. Others say that countries with easier regulatory pathways—like Hong Kong, Singapore, and Japan—are the fastest route. It will be important for the FDA to review the potential for contamination by bacteria, viruses, and other biological agents. As a strapped governmental agency, will the FDA have enough resources to visit these plants and a thorough understanding of the technology to assess it? "Unless you have a perfectly sterile facility, with a cleanroom, and the bioreactors are being operated by robots, you're at risk of some kind of contamination," Wurgaft said. On the plus side, highly controlled bioreactors can be screened in real time, and could even include remote assessment with all the data in the cloud. In comparison, access is heavily guarded on factory farms; and worker safety in meat-packing plants is failing, as evidenced by the many outbreaks that arose during COVID-19. We think there is consumer protection in our food industry, both regulatory and legal, but will officials think of the possible problems before they arise?

Even experts at the FDA, which oversees tissue growth for medical uses, consider growing food-grade meat a knotty topic. In a public hearing on using animal cell culture technology for producing food, FDA consumer safety officer Jeremiah Fasano said that even if traditional and lab meat were identical, there are safety concerns such as different secondary constituents—substances or chemicals that aid in growth and are specific to the organism—to be concerned about, and unintended metabolites—the intermediates, by-products, and end products of cellular "respiration." Living, growing cells produce a host of metabolites. "I think it's fair to say that biological production systems are fairly complex," Fasano said.

Another expert at that hearing was Dr. Paul Mozdziak, a professor of poultry science at North Carolina State University. He talked about the challenges of scaling up, echoing Wurgaft's concerns: "Each place in the [lab] transfer is a place for contamination to get in, bacterial, microbial, viral contamination," he said. The temptation here will be to use antibiotics, which are highly frowned upon in animal agriculture, and many experts say are of greater concern to public health than climate change. In high-tech environments where safety regulations must be followed to the letter, much depends on how careful the employees are. "In cell culture, most contamination comes down to personnel. Somebody did something wrong somewhere, and it tends to be very difficult to trace," he said.

Beyond contamination, there's another area to ponder. These cells are cloned, and cloning in the trillions leads to genetic differences. Not all the time, but it's there. In *Billion Dollar Burger* (2020), Mark Post tells the author Chase Purdy that this is a possible "hazard." With every replication that DNA is "sensitive to genetic mutation," he said. This leads to unstable cells, and presents as a challenge for startups as they make plans to churn out fake meat. Eating these genetically altered cells, we're told, isn't a threat to our health. We don't actually know if this is true.

Fears of safety, viral outbreaks, and "Frankenfood" may be outweighed by calamities that push us toward adopting cell-based meat. In 2001, an outbreak of foot and mouth disease in Britain caused more than six million cows and sheep to be culled and burned. In 2019, African swine fever hit China, decimating the biggest herd of hogs in the world. At the end of that year, coronavirus broke out first in China and then spread rapidly around the world. It was reported to have begun in live meat markets, but many blamed our continued spread into fragile ecosystems, claiming ever more land to raise animals for consumption. As 2020 wrapped, the COVID-19 pandemic was still keeping us indoors and altering our everything.

DIY cell-based meat can't be set up in your kitchen like pickling and canning. Our food is evolving, it has evolved, light-years ahead of our understanding. Wouldn't we all be better off eating tofu and tempeh? Frequent consumption of red meat is widely thought to be detrimental to our health. Some studies have pointed to the heme found in today's industrial beef as cancer-causing. The heme found in grass-fed beef is not identical, although there are no conclusive studies yet about its effect. Will eating cultured meat daily be equally as bad for our health, or can scientists tune the cells to reduce saturated fats and cholesterol? If we embrace animal meat made in giant bioreactors, are we giving up the right to know all that went into its creation? The questions here are endless.

What's the Verdict?

"I think most consumers are just going to look for the thing that is most delicious," said Bruce Friedrich, looking at me over a giant conference room table in a fancy hotel in downtown San Francisco. "If they like the taste and the price is reasonable, I don't think there are a lot of dairy consumers who are going to insist that cows are milked." This was September 2019 and Friedrich and I were together at GFI's annual conference for plant- and cell-based meat. With previous managing roles at both PETA and Farm Sanctuary, Friedrich is a staunch animal rights activist, but despite his feet being firmly planted on the side of animal protection, big names in meat—Purdue, Tyson, and JBS—are now welcomed at his conference. However, their wares were not shared. For two days the attendees of Friedrich's conference dined solely on plant-based treats. I ate everything, including a delicious fried chicken that had realistic muscle-y fibers that pulled apart convincingly. It was made by Worthington Foods, which I wrote about in Chapter 3, and was one of the very first companies to tackle turning plants into meat.

Delicious is helpful, but in a study by Peter Slade from the University of Saskatchewan , Slade posed this hypothetical question: Which burger would you buy if they all tasted the same and were the same price? The beef burger was picked by 65 percent; the plant-based burger by 21 percent; the cultured meat burger by 11 percent; and 4 percent of consumers would make no purchase at all.

This is all fodder for projections and guesstimates. What's sure to happen at the start is that cultured meat will be sold at a premium price to those who can afford it. Industrial meat will continue to be pushed downward to the demographics that can't afford better—the same unequal structure we have today. The same communities that are being hit hardest by illness and disease.

There are other arguments for being more thoughtful to how cultured meat is introduced to the world. In an article in *Slate,* future innovation expert Spackman wrote that cultured meat "will continue disrupting the metabolic intimacy located in a first-hand experiential understanding of where food comes from." Our fascination with how technology can solve food shortages and environmental degradation comes at unknown costs. Cultured meat can't just cancel animals completely out of the picture. Spackman wrote: "This is where the logic has failed us. The cow does exist and participates in a cycle and has an immune system and is part of a chain that is renewing the earth and not feeding the people," she said.

It's also possible that cell-based meat could go the way of the hippopotamus, which was pitched as a solution to a looming protein crisis in the early twentieth century. In the December 28, 2013, issue of *Wired,* Jon Mooallem, the author of *American Hippopotamus,* recounted that there were other "out-there ideas" to answer the meat shortage—including importing antelope and building ostrich farms. "They were basically open to everything. But you'd end up with a constellation of local food systems—a very 'Michael Pollan-esque idea,'" said Mooallem.

That isn't how it played out. We didn't get hippos. What we got was industrial agriculture focused on a very limited subset of animals. It ushered in cheap meat, which led to our increased appetite for it. That in turn transformed small farms rich in biodiversity into mega farms growing few crops—mostly corn, soy, wheat, pigs, chickens, and cows.

While we wait for cultured meat to launch, there are things that can help recover what we have lost: Do better with the farmland we're already growing on. Improve the soil, support regenerative farming, focus on biodiversity and crop resiliency. If you eat meat, grass-fed, pastured meat supports a local and regional regenerative farming family. Bonus points if you buy from one that practices carbon sequestration on their grasslands. Put the best minds together to use our resources more efficiently. Reduce food waste—get food to the people who need it instead of throwing it away. Last, shift our diets from meat-centric to plant-forward.

Valeti has said on record that he'd like to offer tours when Memphis Meats opens its production facility. As a foodie, anytime I can get behind the scenes and peer into food production, I'm in. I love factory tours, and recall going to the Blue Diamond almond factory with my dad when I was very young. The machinery and the assembly line made a wonderland of efficiency. This is different. Imagine standing there with your inquisitive eight-year-old trying to explain the massive steel tanks and what's in them. It could be an exciting conversation—scientists have figured out a way to grow meat in tanks! Or it could be a message of warning—many years ago we used to eat animals. Maybe, next door to the factory, Memphis Meats can have a tiny farm for kids to visit. They can moo like the cow, cluck like the chicken, and quack like the duck. Of course, someday we might also have to explain to our children what a farm was—that once, long ago, we used to grow our food outdoors on the land.

Chapter 9

Are We Buying What They're Selling?

Profit or Health?

For Suzy Amis Cameron, environmental activist, mother of five, and wife of film director James Cameron, the path is clear. Eight years ago, after watching the documentary *Forks Over Knives*, the Camerons switched to a plant-based diet overnight. Later, they divested of all investments that were antithetical to their new lifestyle. This included closing a dairy in New Zealand, a country known for its dairy exports, and turning it into an organic farm. "Deep down inside I wish everyone was vegan—it would be better for the environment. It's just a win-win-win all around," she told me. (Yes, there are a lot of vegans in this book.)

Cameron and I met at an event called Future of Food 2.0. Plant-based startups passed out samples inside a packed WeWork location by the San Francisco waterfront. Afterward, Cameron sat on a panel about plant-based investing. In addition to the farm in New Zealand, the couple invested in a facility to grow, harvest, and process beans, lentils, and peas in Saskatchewan, Canada. As a vehicle for getting peas and lentils into more people's hands, Cameron created a food company called One Meal A Day (OMD). In Cameron's eyes, it was mind over matter. If everyone in the United States ate one plant-based meal a day for a year, it would be "the equivalent of taking twenty-seven million cars off the road," she said. Saving the planet was imperative. "We all have to realize that we have to do something for the environment. It doesn't matter if we drive an electric car or have a nice home or are healthy, if we don't have a planet to live on."

This struggle over which "thing" was more important, our health or our planet, arises repeatedly. One organization working to get the message out that they are equally important is the World Wildlife Fund. In a new report, the nonprofit puts forth how "dietary shifts in countries around the world can bend the curve on the negative impacts of the food system." Changes to our diet, and the way we grow, handle, and distribute our food, will improve our health and our planet's health. On the topic of how technology might shift what we eat, Brent Loken, WWF's global food lead scientist, thinks it all has potential. "I'm just not sure how much potential," he said. "I'm skeptical they can be scaled in a way to have a huge impact. But . . . you never know. With cultured meat, we have to look at it from a health and environmental perspective. We can't say cultured meat is more biodiversity friendly and creates less GHG. If it's not good for your health, then that's a big problem. We see a lot of companies touting plant-based burgers, but they're not necessarily healthier than real [meat] burgers. They have to serve both."

I want this, too. But human health is my priority. It's too important and too often relegated to a line item that is subservient to profits. Because I have type 1 diabetes, I must be vigilant in understanding what exactly is in my food. But I'm human. When I slack off, cheat, or fall off the low-carb, minimally processed cart, I experience the physical consequences almost immediately. Despite these complexities, I can afford to shop at the farmers' market and cook healthy meals at home. When I have questions, I can email my endocrinologist. My privilege is clear.

How different it is for so many working Americans who struggle to afford healthy food, cook at home, or find time to read endless studies. These people face life-threatening consequences—such as obesity, cardiovascular disease, diabetes, and hypertension—caused by structural inequities endemic to our food system.

The primary objective of the food industry is to sell stuff, not to provide healthy foods. We do well to remember this when we walk through the supermarket, tap on an Instagram ad, or fill an online shopping cart. While the status quo may have subtly evolved, including attempts to reformulate ingredient labels—removing additives and synthetic food dyes, lowering sugar and sodium—unhealthy food additives are still the status quo. Case in point: We're still swamped by a deluge of soda, candy, and snack foods, years after everybody agreed they're bad for us, and the marketing continues to prey on our hardest-hit communities.

When I set out to write this book, I knew I wanted to spend time considering how new companies were bringing their foods to market, what the messaging would be, and, since profits were some distance beyond the horizon, what were their (other) primary goals? Are these New Foods being created to save the planet, save the animals, or save us from our bad dietary choices, which are themselves the result of Big Food?

In tandem with lofty aspirations, New Food startups need to make their product delicious, convenient, and cheap. Want extra credit? Make it clean label with fewer, simpler ingredients. To deliver food at a price everyone can afford, or sell ingredients to companies that are competitively cheap, is challenging. Similar to the drug industry, only with *less regulatory oversight*, New Food is investing millions in research and development. Once they've done the heavy lifting of creating their product and selling it for the same as or less than its analogue version means startups must scale up and match a commodity market that has had decades to establish itself.

But what is it really all about: climate, animals, or us? Most, if not all, of the experts I spoke with invariably wound up settling on the climate crisis, which is directly related to our food system. "The issue is only that we eat too much meat," said Christy Spackman, the future innovation expert from ASU. To get that trend to shift, we'd have to fundamentally change how we're approaching the problem, she said. "We've thrown a technology fix at a problem we've created, rather than addressing the [underlying] problem." Easier said than done. The story that cell-based meat makers are telling us, said Spackman, is that killing animals is wrong. "That argument works in *that* value system," she said. Saving the planet through lab-grown meat has many more problems—breaking foods down to their molecular composition, thinking that we can remove extraneous material, and it will have no adverse effect. But there's a good story to tell, and to sell. "This enables more people to make money," she said.

Save the environment, stop the slaughter of animals, and make a profit for already rich investors. What this means is that we're on our own: read nutrition labels, understand how food is made, and eat with your health in mind.

The Fermentation Narrative

Much of the flavor in our foods is microbial. In the yeast family alone there are about 1,500 species in the world, yet only a small amount are utilized by the food and beverage industry. We celebrate their use in beer, wine, bread, and cheese, but beyond that the term "microbe" makes us think of diseases and germs, and the 2020 coronavirus pandemic may serve to heighten our fear of invisible bugs. Microbiologist Dr. Anne Madden wants to change this mindset. "My mission in life is to reveal the utility of the microorganism around us," she said. She found her first commercial yeast in the belly of a wasp. Weird, yes, but it's already being used to make sour beer, cider, and sake. That kind of weird I can handle.

The language around what Madden is doing is tricky. "Brewers think that 'wild yeast' means hard to work with; consumers think it means natural and good." How companies choose to speak about their food shapes how we consider it, and whether we buy it. "Consumer education" is top of mind for many of the startups in my book. Sure, it would be easier to add the word "artificial" or "imitation" to these new plant-based foods, but both words have negative connotations. "It's not that the origins are odd, it's just that we may not be comfortable speaking to them," said Madden. "The challenge is how do you gracefully educate a public in a way that they can make informed decisions?" While Madden is hoping to discover yeasts that already do what we want, food makers in this book are looking to hack microbes to produce something else. For Madden, one step is using language that doesn't unintentionally mislead.

The word "fermentation," which is a large part of making beer, is one of the cornerstones used by synthetic biology startups to sell their better-than-whatever promises to the world. The marketing is almost plug-and-play, piggybacking on two decades of craft beer hipness and easily capturing younger generations of consumers. "So we're going to put these cells in an artificial environment, and try to make them grow. It's like what you would grow yeast in if you were making beer," said a Memphis Meats scientist in *Meat the Future*, a 2020 documentary by Liz Marshall, a Canadian director whose work focuses on environmental and social issues.

This synthetic-biology-as-brewing platform is vital to the New Foods ecosystem—but unlike beer, where you know what goes into the tank (grains, yeast, hot water, hops, flavors) and what comes out (beer), we don't know much about what goes into making these novel proteins that aren't grown inside an animal's own musculature. The shroud of secrecy around recipes and processes is presented as a necessary evil, and proprietary technology is the come-hither for investors assuring them that the horse they backed can win.

In 2017, when I talked to Pat Brown at Impossible Foods, he told me that filing patents would allow him to "talk freely" about their processes, including how they made their heme, which is not made using the fermentation we're familiar with and celebrate, for example, at a cheese stand at the farmers' market. "We're not a gigantic company," Brown said. "We *are* vulnerable." He was referring to the dozen or so multinational food and beverage companies that control a huge chunk of the market, and which might pounce on his company's hard work. "One of our strategies was that we weren't going to rely on trade secrets," he said, referring to the step-by-step instructions one might need to make an Impossible burger at home. Another nod to Big Food. "If we have a critical form of IP, we're going to patent it, which means as part of that process, for something that has to be patented, it has to be shared." As of late 2019, Impossible had 139 active patent applications; 16 of those had been granted.

For the record, I'm not an investor in any food companies. In my efforts to understand the mindset of a food-tech investor, I sat down with Brian Frank, an investor in Plantible Foods, which I wrote about in the chapter on algae. His food- and ag-based venture FTW Ventures, focuses on companies using "responsible science" to reinvent our food system. This mission-based agenda is a common thread for founders and investors. Almost everyone wants to believe that they're fixing problems, canceling out agendas, and bringing better food to the market. When Wonder Bread was created, it was to give the American public a standardized sandwich bread that could sit on our counter for two weeks and remain soft and squishy. The fact that it was flavorless and had less than 1 gram of fiber didn't matter. In its era, it was embraced.

I met with Frank at a café in downtown San Francisco. It was loud. Seated at every table beside us was another pair of techies sipping cappuccinos and loudly chatting through their own business meeting. The blond and bespectacled Frank talks very fast, and seems to know everyone. His stable of investments at the time included Geltor, Square Roots, and Proper Food. Not a bad list, but as we were in the Bay Area I had to ask: Why didn't you invest in Beyond Meat or Impossible Foods? To make his point, he asked me rhetorically: "Do you want to be in the hot investment?" Answering his own question, he shook his head and continued: "I want to be in the 'It's not hot now, but will be hot next.' I'm going one click beyond what people are [currently] freaking out about." With both companies skyrocketing in valuation, and Beyond Meats' immensely successful IPO, I pressed him to say more. "I would have done Impossible before Beyond because Impossible has science, something that underpins it, but hindsight is 2020. Of course, I would have done both," he said, laughing.

For Impossible and other companies like it, patents on new foods* are a way to present what they're doing as more than the basic sustenance of life. Bringing new and critical upgrades to our lives is part of the promissory narrative of these new food companies. Tomorrow's burger, e.g., Impossible 3.0, which isn't out yet, is something to try, to be downloaded like the latest software update. On the Aleph Farms website, we're told that the company's superior, healthier, more humane, slaughter-free meat will provide "a new customer experience." (Whatever that is.)

Media coverage of future foods is more often complimentary than critical. They are treated like an hors d'oeuvre, or a trip we're planning—something to look forward to. Startups depend on consumers becoming familiarized with their curious new foods, and they look for ways to normalize what they're doing. This is done through the dozens of stories we can read about a new product before a launch or by its visual representation on the web or an app on our phone. Food stylists are hired and photographers are booked. The marketing shots look like every big, juicy hamburger pinup we've ever seen. Remember when Carl's Jr. hired swimsuit models to eat their "sauce-dripping juicy burgers"? Like that, minus the hot bod. A close-up of human hands squeezing pea protein patties and all the fixings between a sesame-seed bun. Cow-less whey protein packed into a pint of delicious-looking ice cream. Cell-based fish cut into cubes for a poké bowl. We tweet about them. We talk about them on social media. We are their pre-market evangelists.

Halfway through researching and writing this book, some of the narratives praising plant-based meat began to shift. Nutrition professionals began writing that Impossible and Beyond's burgers weren't actually healthy, but another example of ultra-processed fast food sending us off to more future cardiovascular disease.

* The company Just has about forty patents, several acquired from Willem van Eelen, one of the earliest scientists to consider lab-based meats as a way to end hunger.

In 2020, right before Labor Day, a holiday known for backyard grilling, a war on ultra-processed escalated in the form of a series of full-page ads taken out in the *New York Times*. The flames were fanned by Lightlife, which sells plant-based proteins such as tempeh, burgers, and hot dogs. The headline: "An Open Letter to Beyond Meat & Impossible Foods." The subhead: Enough. "Enough with the hyper-processed ingredients, GMOs, unnecessary additives and fillers and fake blood," Lightlife wrote. It went on to declare it was making a "clean break from both of you 'food-tech' companies." The "real food company" wrote that "people deserve plant-based protein . . . developed in a kitchen, not a lab." Impossible Foods dashed off a rebuttal on Medium that was written by its communications team.

The ad from Lightlife, they wrote, was a "disingenuous, desperate disinformation campaign attempting to cast doubt on the integrity of our products." Impossible was quick to point out that Lightlife is owned by Maple Leaf Foods*—known more for its packaged meats. Beyond Meat sent its response via an email to the website Food Dive. Unlike Impossible, it did not go on the attack but pointed out that its burger was "made from simple, plant-based ingredients." No GMOs, synthetic additives, carcinogens, hormones, antibiotics, or cholesterol. Marion Nestle, an author and professor of nutrition, and someone who never backs down from a fight, wrote that from her standpoint "the differences between these products are minimal." They don't look like their origins, are industrially produced, and cannot be made in a home kitchen.

The tit for tat continued. In a follow-up, a new contender jumped in with their own full-page ad. Planterra Foods, which has new plant-based burgers being sold in the United States at Kroger, wrote that they'd be "remiss if we didn't take a moment to say thank you to Beyond Meat and Impossible Foods who shined a light on this space and helped elevate it to where it is today." What the Boulder, Colorado, company chose not to

* In addition to Lightlife, the Canadian company, which had almost $4 billion in revenue in 2019, also owns the well-known vegan brand Field Roast.

share was its main ingredient supplier—MycoTechnology, which I covered in Chapter 2—or that it's owned by JBS, the largest meat processor in the world.

There are common assumptions we tend to make about our food these days. Fewer ingredients, less processing, and names we recognize are better. In this equation, does it matter what the motives are of the company that is producing our food? In *Meals to Come*, Warren Belasco wrote, "The food industry profits mainly by concentrating calories into highly processed, value-added steaks and snacks." Although his book was written in 2005, little has changed.

It may be that I have too much Marion Nestle in me. In her numerous books on our food system, Nestle is critical about our knowledge of nutrition and especially about Big Food, which includes these startups. I can't help but hear her voice in my mind: "I would say that people should always be skeptical of breakthroughs—if something seems magical, then it's probably not real. For example, there's no such thing as a superfood."

The Mayo Story

In 2013, San Francisco-based startup Eat Just, formerly known as Hampton Creek, launched its first product, an eggless, plant-based mayo. The mayonnaise knockoff was greeted with acclaim. It was as if the world had never seen a condiment before. They wrote: "Founder Josh Tetrick is out to change the world, starting with mayonnaise." And that "it took two years of research and development to figure out how to make mayonnaise without using eggs." A wave of stories covered the then thirty-three-year-old Tetrick. Journalists wrote about the product, the investment money, and the awe-inspiring technological advances the startup used in the name of sustainability.

The problem was, this product already existed. Vegenaise—a mashup of the words "vegan" and "mayonnaise"—was first developed in the mid-1970s by Follow Your Heart in San Fernando Valley, California. Before becoming the vegan product powerhouse it is today, Follow Your

Heart was a natural foods market with a cozy twenty-two-person vegetarian café inside. One of the staples in its kitchen pantry in 1974 was Jack Patton's Lecinaise, made from soy lecithin. Bob Goldberg, co-founder and CEO of Follow Your Heart, used it on everything. He called it his "secret ingredient." At some point a rumor circulated that there were eggs in this supposedly eggless mayo. Goldberg reached out to Jack Patton, the owner of Lecinaise, who assured him that it was egg-, preservative-, and sugar-free. Goldberg was reassured. The California Department of Food and Agriculture was not. The agency raided Patton's Lecinaise facilities and found workers soaking the labels off regular mayonnaise to use and sell under their own brand name.

Goldberg was floored. His café depended upon this product. He looked to other manufacturers for help. "They all insisted that there was no way to make mayo without eggs," he said. In his home kitchen, he blended up batch after batch trying to achieve the taste and consistency. Finally, he nailed it by combining almond oil and tofu scraps. Unfortunately, when they first brought it to market in 1977 grocers didn't stock it in the refrigerated section and the oils separated. Follow Your Heart took it off the shelf, and made plans to relaunch it when it could solve the issue. They took their time. In 1988, with its own facility, it switched to canola oil, and prayed that consumers would be okay with a refrigerated product. They were, and are: It's still Follow Your Heart's number-one seller, available in ten varieties and dozens of countries.

The difference between Tetrick's mayo and Vegenaise is small. Tetrick made his shelf stable so that it could safely sit on the shelf for six months. Vegenaise, with fewer ingredients, must remain cold. Products that are shelf stable require gels, gums, or stabilizers. Tetrick used modified food starch. But Just wasn't even the first here. Hellmann's, Best Foods, and Kraft have been making shelf-stable mayo for decades. This leaves us to wonder why exactly the world went bananas for a condiment that was already on our shelves?

Tetrick, who grew up in Alabama, speaks with a deep Southern drawl that stands out in Silicon Valley. It worked magic with the press. He spun a yarn. He gave good story. He proclaimed that he was going to "make the egg obsolete." Do not underestimate the power of his boast. Tetrick was one of the first founders to propose ending our reliance on animals for their protein. He explained to journalists that the ratio of energy input to food energy output for a chicken egg was a lot higher than the energy needed to grow a crop of plants. This quantified sustainability. It was an alluring message and the media jumped on it. Just's big-name investors, including Bill Gates, Peter Thiel, and Khosla Ventures, helped. With social media exploding, it was as if Vegenaise had never existed.

But it did, and Goldberg, its inventor, is an all-around-good guy who was fine with being out of the spotlight. More mayo sales for anyone, he said, would eventually mean more sales for his own products. Goldberg is the epitome of LA casual. Now past seventy, he wears a uniform of shorts and sandals and sports a long, gray ponytail. The only flashy thing about him is his red Tesla with the license plate VEGNASE. Goldberg doesn't hold media events. In fact, he didn't hire PR until around the time of Just's mayonnaise launch. When it did, Follow Your Heart went out with its own story. "We couldn't afford not to," Goldberg told me in one of our many conversations over the years. "It felt like a personal affront. Every time we saw an article about this amazing new incredible mayo without eggs, we were incredulous that no one would even Google to find out that [our] product was out there for decades," he said.

These days, anything "new" gets a story. *Forbes* magazine covered Tetrick's mayo launch at Whole Foods Markets in 2013. They preached about the startup. They wrote: "Hampton Creek has examined the molecular properties of 1,500 types of plants to find species with the best characteristics for emulsifying into mayo or congealing in a hot pan like scrambled eggs." One look at the ingredient label and you'll see that Just also used canola oil in its formulation. "I was told by someone from the company that they had jars [of Vegenaise] all over," said Goldberg.

The frenzy might have subsided, but there was another twist. In 2014, Unilever (owner of top-selling Hellmann's mayonnaise) filed a lawsuit against the startup for using the word "mayo" on its label. In 2016, Tetrick was accused of artificially inflating sales by instructing his employees to buy up their own product. Shortly after this damaging news, Target dropped the brand in all of its stores because of supposed food safety concerns. As for the lawsuit, Unilever eventually dropped it because of a backlash of bad press. The FDA, however, was still on the case. The agency eventually came to an agreement with Just to relabel its mayo, removing an illustration of an egg and adding the words "salad dressing." In 2017, in an attempt to distance itself from the bad press, Tetrick's company changed its name from the rustic Hampton Creek to the simpler (but more difficult to use in a sentence) Eat Just.

Food innovation and its subsequent copycats are now something we need to account for when we go to the supermarket. Garrett Oliver, head brewer at Brooklyn Brewery, told me that when he bought a jar of Just Mayo a year ago he "was absolutely livid" to discover that it was egg-free. His anger was twofold: It was called "mayo," and it was called "Just Mayo." Oliver has spent a lot of time thinking about how foods have changed over time. He often uses beer to tell his stories. "The basic arc of the idea here is looking through beer at the evolution from reality, which kind of existed in the 1800s into the early 1900s, and then the scientific transformation of food from the real thing into simulacrum." Oliver calls this "the matrix." In beer it was when big brewers wanted a drink that tasted the same everywhere in the world. Beer, which depends on evolving crops and microbes to turn grains into alcohol, was once a vast and varied product that became a mostly consistent, flavorless liquid, only to return once again to the land of highly divergent craft beer. The supermarket from Oliver's childhood, which had bread that didn't taste like bread, and only four kinds of cheeses, "was one gigantic lie. Every element in it was false," he said. "If you can make people forget reality, then you can replace reality with something else." Were the falsehoods on our shelves in the last century—Wonder Bread, margarine, Velveeta cheese—any different from a creamy, eggless spread called Just Mayo?

It took more than four years before the release of Just Egg, a liquid egg scramble made primarily from mung bean protein. Tetrick relied on the same strategy: press outreach before the product was sold commercially. He invited journalists (myself included) to try the "eggs," and had his team of Michelin-star chefs whip up omelets. He told me: "On the taste front, it doesn't need to taste like a conventional egg, but the creamiest. We want it to be better than a farm fresh egg." The omelets were yellow like eggs, and they were indeed creamy and delicious. But they were not eggs, which I was fine with. Why not give them a new name? This time, if Tetrick made any missteps, they didn't go public, and Just Egg has managed to gain mainstream appeal and impressive sales. By the end of 2020, the company reported to have sold the equivalent of seventy million eggs.

"Marketing should be about making people aware of your products in an honest way. Without hyperbole," said Goldberg. That these founders are different is obvious. Tetrick is only doing what others like him are doing. Tell your story loud and make it feel original or be first of its kind (even if it isn't). Get in front of journalists early and win them over. Raise money from investors and then put it into making your story a reality.

When I met with Tetrick in his open-plan office in San Francisco he told me: "If we focus on what we described, people will choose us instead of beef and tofu." Describe it, and then make it happen. Demand for chicken eggs shows no sign of abating, and the faux market is exploding with options. Our food choices are becoming ever more complex, and no one has time to make a thoroughly informed decision by plowing through the glut of information, the claims and counterclaims.

A few weeks after I talked to Garrett Oliver in Brooklyn, he emailed me a photo he'd taken in his local Brooklyn supermarket. In it: a bottle of Just Egg next to a carton of real liquid egg whites. He wrote: "Look at the color and size of the word 'egg' next to the size of the smaller print above?" In tiny lettering at the top of the bottle it read: "Made from plants (not chickens)." "Completely, utterly fraudulent," he wrote. Retail data from supermarkets shows that placing plant-based products next to their real counterparts drives increased sales to plant-based. Is it because consumers are grabbing the wrong thing? I find myself increasingly reaching for Just Egg because I like it, and prefer to be eating more plants, but it was hard to argue with Oliver's point.

Real Versus Fake

Naming the next generation of bio-similar foods will be convoluted. Where once the branding of novel foods was relegated to a private group of experts, it's now being brought into the public arena. The opinions are strong, and include the FDA, lobbying groups, food associations, legal bodies (even the ACLU), and entrepreneurs. An example: In January 2019, the FDA asked for consumer feedback on the use of classic dairy names by similar plant-based versions. Consumers have the opportunity to voice their two cents' worth, but the outreach is seen by few. If we're honest here, not that many of us take the time to provide our comments and concerns beyond shaking our fist in the air and moving on with our day. As for the FDA's 2019 request for comments, no decision has yet been handed down.

By the time the FDA jumped in, there were already dozens of dairy alternatives on the market, and happy consumers buying them. Soymilk has been a common drink since the 1980s. So why the fuss now?

One month after the FDA's request for comments, Jasmine Brown filed a lawsuit against popular plant-based cheesemaker Miyoko's Creamery. Brown, most likely a name fronting for a money-hunting lawyer or a dairy industry group, outlined in the class-action complaint that marketing "butter" made from cashew cream was confusing and misleading. The attorney wrote that Miyoko's Creamery inaccurately used the word "dairy" in its labeling, its packaging looks like butter with a yellow color band on the box (yellow means butter!), and it states that the product "melts, browns, bakes, and spreads like butter." Consumers may believe the product is as good as real butter and pay $6.99 per package for it, the complaint stated. What it didn't say was this: Vegan butter now tastes as good (or almost as good) as the real thing. Traditional commodity categories—butter, milk, cheese, and yogurt—are hanging on, but their hold is tenuous. Now that our grocery stores are stocked with a whole range of competing products, legacy makers are doing whatever they can to thwart sales.

In addition to issues with naming, the suit against Miyoko's claimed that cultured butter made from cashews was not nutritionally equivalent to butter made from cow's milk. If you're curious, Miyoko's butter is made primarily of coconut and sunflower oil. It does include some cashew cream. A single serving is 100 percent fat, and the cashews give it a dose of magnesium and iron. One serving of butter made from cow's milk is a close match—all fat plus a minor amount of vitamin A.

Understandably, the dairy industry is scared. Fluid milk sales have plummeted, large commercial dairies are closing, some are filing bankruptcy, and profit margins are getting even thinner. In the non-dairy section of the refrigerator, the brands and options seem almost infinite. Switching to plant-based milks is often the first step to shifting to a plant-based diet. Who can argue with marketing slogans like "Better for you, better for the environment," and "For our bodies, our planet and our future"? In a 2020 study on the politics of plant-based milks, Sexton, the human geographer at Oxford, refers to this as a "palatable disruption," and that "people are encouraged to care about the environment, health, and animal welfare enough to adopt [plant-based milks] but to ultimately remain consumers of a commodity food."

As a disruptor, plant-based milk is not comparable to vegan cheeses, which are tasty, but are in no way a threat to traditional cheese. In the plant-based milk category we can already see the machinations of a mature industry from launch to marketing to consumer acceptance and ingredient evolution. Supporting Sexton's work is an idea she borrows from sociologist Jesse Goldstein's book *Planetary Improvement* (2018) about nondisruptive disruptions, which are "technologies that can deliver 'solutions' without actually changing much of what causes the underlying problems." Sexton points to Danone, a multinational that owns both plant-based brands and a huge dairy portfolio. In 2018, it made $1.9 billion in plant-based beverage sales, and by 2023 it promises to triple that. Given this, will Danone* stop making dairy versions of our staples? I doubt it.

While the dairy industry shifts its priorities—making cheese, whey, and yogurt or buying plant-based companies in order to hedge their bets—meat lobbyists are busy working to ensure that plant-based meat companies can't use the word "meat." The first state to enshrine a traditional definition of animal meat in its laws was Missouri, which passed a bill in May 2018. The definition for meat outlined in the bill was "any edible portion of livestock or poultry carcass or part thereof" and prohibits individuals from "misrepresenting a product as meat that is not derived from harvested production livestock or poultry."

The state representatives backing the bill hoped that their Real MEAT Act, which stands for Marketing Edible Artificial Truthfully, would provide grounds for the USDA to enforce the new rules, if the FDA didn't take action against these "confusing protein claims." (The FDA regulates 80 percent of America's food supply, but it's the USDA that oversees beef, chicken, and pork.) Several states have followed Missouri, passing laws restricting plant-based or lab-grown proteins from using terms such as "beef" or "meat." Not willing to sit back and watch, Tofurky—with legal aid from the ACLU and GFI—sued the state of Arkansas over a law that would prohibit the forty-year-old company from using words like "veggie burger" and "tofu dog" on product packaging. (Why hasn't some animal lover sued every hot dog maker for using the word "dog"?)

In their rebuttal to this flurry of legal activity, the Center for Science in the Public Interest (CSPI) fired off a letter to the USDA Food Safety and Inspection Service. They wrote that restricting the use of the words "meat" or "beef" to the traditional animal product was "unnecessary to

* In February 2021, Danone announced it would acquire 100 percent of the shares of Earth Island, the parent company of Follow Your Heart.

avoid consumer confusion," and "rather than serving consumers, the petition represents a self-interested attempt to restrict healthy competition between industries vying for space at the center of the American plate." Any jurisdiction over labeling, they wrote, should take into account the entire context of the label. But the USDA has a history of favoring a narrow band of commercial interests, and it's unclear how they will respond to the advent of these new industry players that are much savvier in gaining consumer advocates. As for Miyoko's, the company filed a lawsuit against the California Department of Food and Agriculture in February 2020 on the grounds of freedom of speech.

The Big Boys' Marketing Playbook

Food tech startups are borrowing their marketing strategy notes from the big guys: Nestlé, Kellogg's, General Foods, Tyson, PepsiCo, and Coca-Cola. The product launch and brand positioning are crucial. These startups call their creations "plant-based," not "multi-ingredient processed non-meat/dairy." Much like the meaningless term "natural," "plant-based" can shield products that are not actually that good-for-you behind an apparent halo of good health. If you think about it, Coke can be considered plant based because it's made from plants—corn syrup and cane sugar.

When biochemist T. Colin Campbell coined the phrase "plant-based" forty years ago, it meant something different. At the time, Campbell was part of a team investigating the connection between cancer and nutrition. The vegetarian diet back then was so far outside the norm that Campbell felt the term "plant-based" would have fewer negative associations. His later book, *The China Study* (2004), was a seminal work on the benefits of a vegetarian diet. In it, he added the words "whole foods," as in whole, plant-based foods, in order to avoid implying that isolated nutrients—such as supplements or plant-food fragments—conveyed health.

This point is big, even today, as greater numbers are converting to more healthful diets. Food packaging still touts specific nutrients, vitamins, and other sources of better-than whatever in order to get our dollars. Novel foods may look and sound healthier, but they're still a fabrication of stuff, not a whole, plant-based food.

It's debatable whether food-tech founders have used science to build a better ice cream, or a burger that has less impact on the environment, but look at how far they've come in just a few years. Beyond Meat is sold at almost every major grocer—Whole Foods, Walmart, Kroger, Safeway, Target, and more. Impossible is sold in more than eight thousand retail stores and seventeen thousand restaurants, and you can now order it online direct from the startup. In the conference room at the Impossible offices, Pat Brown announced to me: "We are not only going to be the most impactful in terms of future-of-plants company in history, we are going to be the most financially successful company in history." When he said this, I thought he sounded like an evil genius. I thought of robber barons and Wall Street excess, not a scientist hoping to improve our world. He continued on his soapbox. "We're going to make [our investors] more ridiculously rich than they already are." By 2030, the anticipated market value of meat will be $3 trillion.

Price doesn't seem to be a sticking point. You can order a regular Whopper for about $5.99, or pay a dollar more for the Impossible version. (At fancier burger spots the burger goes for $18 to $22.) Someone is making money, and fast-food restaurants are almost certainly benefiting from increased foot traffic. While Beyond Meat chose the supermarket route, Impossible Foods initially worked exclusively with chefs, hoping to benefit from their credibility. Impossible also donates hundreds of pounds of ground Impossible meat—one can assume it's the reject plant meat—to food banks in Alameda and Santa Clara County, and sent chefs to teach the food centers how to work with the product in a "fun training session." In addition to fast food, getting placed in food pantries offered Impossible early access to the food insecure, which it further cemented when it launched into fast-food chains.

In order to achieve Impossible Foods' goal to replace animals in meat production by 2035, what's needed is massive, wide-scale adoption.

"We're starting with a bunch of ingredients that have nothing to do with a burger and we're choosing a target to aim at," said Brown. Since launching the burger, Impossible has launched a breakfast sausage, and in October 2020 the company showcased a prototype for its plant-based milk. Chicken, steak, and seafood are in the wings.

Like dominoes, these companies are quickly setting up in Asia, India, and Africa. In China, where food safety is a major public health concern, American brands are seen as superior to domestic brands and imports are eagerly anticipated. While Impossible is looking for local partners to help with manufacturing in China, this timeline may be affected by the coronavirus pandemic. Beyond is already selling out in restaurants and retail locations in Hong Kong, according to its distributor.

Now that we can buy ground plant-based "beef" in the grocery store, purchase it online, and order a vegan-ish burger in thousands of quick-service restaurants, the momentum seems unstoppable. In 2018, White Castle became the first fast-food chain to serve the Impossible burger in the United States. In 2019, Burger King, which has more than seven thousand locations in the States, announced its own plans to serve the Impossible burger on April Fools' Day. The "joke" was on their diners, who were served Impossible's plant patty instead of the red meat they thought they had ordered. Incredulous reactions were shared in a video on YouTube. Pat Brown enjoyed the meta joke. "People will get a burger that they will actually believe is made from an animal, and be told it's made from plants, and think it's an April Fools' joke—and it's not!"

Due to the proliferation of the likes of McDonald's, Burger King, and Carl's Jr., 36 percent of the world eats at least one fast-food meal a day. At each of these chains there is now the option of ordering the Impossible or the Beyond burger, or both. Fast-food restaurants were initially pitched to us as successful innovation and family-friendly fun. Would we still cheer on that same innovation today?

These chains aren't supporting a healthy diet, and they aren't promoting plants. They're seeking new customers, or a way to bring in existing customers with the promise of something new. In the United

States, fast-food chains were the last step toward mainstream adoption of the Impossible burger. First they got the foodies on board. Then they got moms and health nuts on their side by launching it in "natural" markets. Now they've got middle America on board. It's a not-so-subtle operation of infiltrating Americana via its highways and its Happy Meals that is one more reason *to* want to know more. Fast-food chains are home to cheap, quick food that fuels our culture's rise in health-related issues like cardiovascular disease, diabetes, and obesity. It's possible that the fact that I get lumped into that group because of my own condition makes me carry the torch a little higher and louder. Like me, think before you eat. Don't believe the hype.

Chapter 10

What Are We Eating in Twenty Years?

Predicting the future is a fool's errand, yet here I am bringing you an entire chapter dedicated to the question: What's on our dinner plate in 2041?* Twenty years is a preposterously short span of time when it comes to observing real systemic change in what we eat. Even fifty years isn't enough time for that kind of task, which is what Winston Churchill addressed in a group of essays from 1931 called *Thoughts and Adventures* that included the topic of food. In it he imagined a time to come when we didn't raise "the absurdity of a whole chicken," but simply the cuts we wanted—breast, leg, thigh—by "growing [them] separately under a suitable medium." He also, correctly, projected that "microbes would be made to work under controlled conditions" just like yeast. His fifty-year projection took eighty years to become reality.

* If not for the pandemic, my book would have been out in 2020. Not 2021.

So why twenty years? Because I think that today's technological advances are speeding things up, and change is coming to our foods at a faster clip. Because 50 or 100 or 150 seemed too out there. Because Beyond Meat retooled the veggie burger in seven years, and Impossible created its version in five.

Churchill's synthetic chicken isn't what Chef David Nayfeld is looking forward to. Nayfeld, whose own prediction is below, suggested that it was time we learned to appreciate different parts of the animal. "The problem with meat is, we only eat one thing. It's an egotistical view of the Earth," he said. "A cow only has so many ribs, hanger steak, tongues . . . we should eat all of that." I'm in the eat-less-meat camp, but if I'm going to eat it, I'll spend the money on steaks from regenerative farmers. I know this is a luxury many don't have, but the reality is industrialized food startups working to create cultured meat, and other New Food makers, are on a trajectory to become "industrialized." And the question is, is industrialized food good for us humans?

The years I spent digging into New Foods—the startups and their creations—were an enlightening journey. I sampled and cooked my way through almost everything in this book. My mind is open to these ideas and novel products, but many won't succeed, or they will pivot to the new new. The questions posed in this book are important, and figuring out how to prioritize them is critical to our future good health. Do we eat to save the planet, the animals, or ourselves? What about traditional foods from cultures already threatened by a food system that doesn't serve their basic needs? The reasons behind the coronavirus pandemic might point us to prioritizing an end or reduction in eating animals raised in industrial feedlots. This also relies on an end to encroachment on wild spaces for raising more animal meat for human consumption. It speaks to creating a world that no longer spreads into every nook and cranny of Mother Nature. When there's another pandemic, what will our food look like? Will we be any healthier, and who will be in the position to make choices?

But industrial agriculture is a highly efficient way to feed our world; it's the devil we know. Can we return great swaths of land to better forms of farming? Can we feed more people on less land? Chef Sean Sherman, aka the Sioux Chef, votes strongly for learning the lessons of his ancestors, people with thousands of years of ecological knowledge. "We can produce more food if we landscape for it like indigenous communities."

In 1978, science writer Barbara Ford wrote Americans were eating far more protein than they needed—about twice as much. Grain-fed beef may become a rarity in 2000, Ford wrote, with most beef being reared to market size on forage. Her folly here is that she thought the price of beef would go up. It didn't. Cheap corn and soy supported the creation of industrial feedlots and with them cheap meat for everyone that wanted it.

Ford's book shared the "hot" new proteins of her time. These included the winged bean, a unique plant that was completely edible. With seeds and tubers that were 20 percent protein, Ford said it was "almost too good to be true." Then there was the buffalo gourd, a drought-resistant plant that could go a year without water. A year!

As with Ford, speculating on the future leaves me open to missteps. Some foods included in this book could bubble up, then fade away. Other foods that could become staples didn't get in these pages. Namely insects. Like algae, insects don't scale easily, but unlike algae, insects appear to be one of those foods that will remain a part of the cultures that already eat them, but not spread much further. In an interview, Michael Pollan suggested that bugs could be used to feed livestock. This is actually happening, and perhaps I was wrong to think mealworms weren't big enough to make the book. Ynsect, a French company that has raised more than $400 million dollars, is building a factory that will produce 100,000 tons annually of insect protein that will go into fish and pet food. The mealworms' manure will be turned into fertilizer. Can it make the leap to mainstream adoption by humans? Adoption, yes. Mainstream? Not everything deserves to become the next soybean.

Whether insects or algae or pea milk, Wall Street no longer sees New Food as an uncertain investment. At the 1939 World's Fair in New York, Gerald Wendt, a chemist and the fair's science director, said that synthetic foods would initially mimic the plants and animals we already ate. Within two or three generations, he said, foods would abandon all pretense of imitating nature. My niece and nephew, both Generation Z, and raised as vegetarians, may be the ones to cast my doubts to the wind, posting images of their as-yet-invented dinners to whatever social media app is the next thing. If not them, maybe it will be Generation Alpha? Regardless, I'll be old and gray. (I hope.) I'll be eating less food and no meat, drinking less wine, and still complaining about not getting enough exercise.

In the end, this chapter turned into hope. My own hope is that investors will find reasons to direct their vast sums toward bettering the things we already know can feed the world. Let's incentivize farmers to grow a greater variety of beneficial crops that hold the promise of adding to our diet, not taking away. Let's tap the process innovations highlighted in this book to help translate more plants into sustainable and economically positive crops supported by regenerative, local farms around the world. If cultured meat lands on our table, let's use it as a hybrid solution to make plants more delicious, and make a small quantity of better-quality meat go further for more people. Instead of a few unicorns (companies valued at more than $1 billion) every decade, let's aim for a herd.

Now that you've made it this far, I'll step aside so that you can hear from a range of food experts about their own optimistic visions for the future.

Dan Barber, author of *The Third Plate*, chef/co-owner at Blue Hill at Stone Barns, age 51

Seed work on a twenty-year horizon? Let's R & D seeds for the future that are nutrient dense and really flavorful. We ought to select for very regionally adaptive environments, and breed seeds for micro-regions. That's the ticket. The United States is huge and complex. It's not *what* we're eating in the future, it's *where*? That's where restaurants are going. The thing that defines restaurants now is what is local and regionally important—something that you travel for and can't get anywhere else.

What's so interesting is you can get very different ingredients, and you also get a pattern of eating that supports the region.

Food is evolving with investment and technology. I am less enamored with technology investments that want to replace animal agriculture with plant-based meat. It's reductive and doesn't help but clarify for investors what *they* want. Investors seek proprietary IP in order to gain control of the food supply. That's not what I'm for.

I wish for better ecological functioning and better biological understanding. We know how to make a GMO soybean, and make it bleed. We're taking something free in nature, and patenting it. The idea that there are hundreds of millions of dollars [going toward] saving the earth with new foods instead of investing in the foods we know is preposterous.

I'm in pursuit of good food and flavor, and flavor comes from increasing biological systems. The more complex the biological system, the better the flavor, and the more nutrient dense, the better the planet is, and the next crop will be. It's not just getting something delicious, it's: How do you do that over a long time? That's where you need better biological functions. Those are difficult to handle, and impossible to own and capitalize on. That's why business runs away from it.

We have to invest in the kind of farmers and farmland that gives us the nutrient density that our bodies so desperately need as their own vaccine. They're all connected, which is why it's so complicated. It's all one subject. When you talk about vertical farms and Impossible Foods as the answer to feeding the world, you are on the wrong track.

Mark Cuban, entrepreneur, *Shark Tank* television personality, and Dallas Mavericks owner, age 62

In twenty years, I think we'll see the early adoption of synthetic foods—foods that replicate organically grown foods but that originated in a lab. If climate change doesn't accelerate, we won't see cultured meat in ten to twenty years. If we see a further acceleration of climate change with a recognizable impact on weather across the country, then it might happen faster as people realize we may be in deep shit if we don't change how we create and consume food. In twenty years, I think just as we have an index for carbohydrates and other diets, we'll index foods by the impact they have on our environment. The driving factor will be climate change. If it's so bad that even deniers can't dismiss it, then we'll find that index will also incorporate taxes on foods that negatively impact us. This could lead to certain foods becoming so expensive they become rarely available. I hope for a future in food where we have a low-cost cube that is satiating, tastes great, and fulfills the recommended daily allowance (RDA) of nutrients for less than one dollar. This will allow us to end food insecurity.

Kim Severson, food correspondent for the *New York Times*, age 59

There will absolutely be less feedlot meat on our plate. Less meat in general, but there will still be plenty of meat. We'll continue to see processed food that has fewer chemicals. Clean label is here to stay. The frozen aisle will continue to have a transformation. I think people will be able to drill down with a lot more agility on what's better for their bodies.

We're raising a generation of cooks like we've never seen before: canning, making bread . . . Because of technology they'll understand it better, make dishes that are culturally appealing and speak to who they are as well as their nutritional needs. Food will be as fluid to them as digital is, and they'll be as articulate as they are online. They won't be afraid to eat something that's cell-based.

People are hungry for real food. They're going to want to get more in touch with real food, not less. The [cell-based] technology that's developed will go to food processing and manufacturing, and less right off the shelf. Ultimately, people who don't want to eat meat won't eat meat. I think increasingly that people will want more real food.

Class, of course, divides all of this. We're seeing the marriage of the hunger and good food movement. I think the good food movement, doubling of SNAP [food] benefits at farmers' markets, and good nutrition are essential to health. FEMA is being replaced by World Central Kitchen. The food industry was like, "A calorie is just a calorie," but people in need will have fresh and better food.

The stranglehold that big fast-food chains like McDonald's have on how we eat, that will fade. It's already a class division. The younger generation knows that food isn't good. Food manufacturing for specific diets will go away. It's amazing how quickly those products will come and go. I think people will get much better at controlling their diets through real food.

I'd like to see a continuation of food that you want to eat, the food truck mentality, but have it overtake chain restaurants. I'd love to see more local chains. We all need convenience foods, and we need ways to eat when we're working, but really good small chains—a continued celebration of regionality in foods—that would make me so happy.

J. Kenji López-Alt, author of *The Food Lab: Better Home Cooking Through Science*, age 40

Judging by current trends, the average plate will have more meat on it. Meat consumption, particularly among developing countries, is increasing. Even with the advent of plant-based meats and certain portions of the world population lowering their meat consumption, it's not enough to offset the increase in places like India and China.

I think that lab-based meat will become mainstream. It's going to take time. Plant-based meat is still a premium product for now—at fast-food restaurants it's pricier than meat equivalents—but once in-vitro meat is on the market and the cost comes down, it'll start to pick up. It's also a generational thing. Older folks may never come around to it, but, for instance, my daughter knows that Impossible and other plant-based meats are available, and she doesn't think they're weird. I'm sure she'll feel the same way when in-vitro meat comes out on the mass market.

In the distant future, I hope we won't be eating as much meat. The planet just can't sustain it. There's a good chance in a couple hundred years that we'll look back at our past and think, *I can't believe we used to eat meat* in the same way that we look back at, say, smoking indoors and can't believe that we used to think that was OK.

My hopes for the future of food are more political than technological. Because of the nature of capitalism, the value placed on profit over people, and the "haves" and the "have nots," a system of inequality persists. We currently live in a world where we produce more than enough food to feed everyone, yet we still have so many people living in hunger. I hope food distribution becomes more equitable and that government incentives are constructed to emphasize more plants, less meat, and more varied crops.

Ali Bouzari, writer, co-founder of Render, age 33

Anybody who gets too carried away talking about radical paradigm shifts has been spending too much time in Los Angeles, San Francisco, or New York. Meat is a big question, and what role animal products will be playing. If the industry gets more expensive, we'll have fewer animal products because of the economics of it. I wouldn't be surprised [to see] an anti-protein backlash. I wouldn't be surprised to see low-fat as a trend again. I think we're still going to talk about the best version of a Thanksgiving dish, or a hot condiment that's new to us, but not new outside the United States. We'll still be debating the best combo of macronutrients that will make you live forever. I think clean label and clean eating will continue to evolve. People are still going to be eating potatoes, and the individual whole ingredients aren't going to go anywhere in our lifetime. That will be a constant.

I think cellular meat already is in some ways mainstream. The leap of Impossible and Beyond has already made getting people en masse to eat chicken nuggets and burgers easy. Cultured meat is more of a jump in consumer training. I don't think people are going to track lab-based chicken versus plant-based. They'll like brands. Cell-based meat is a generation behind plant-based foods, which have formula and ingredient innovation, but it's still forming and mixing, things we figured out a long time ago. Startups have to figure out how to grow life, but I don't think it's a reach to say in ten to twenty years in the future we could have it in fast-food chains.

I think we'll have gone through more critical evaluations. Any shortcuts to make something that is too good to be true will be reevaluated. Like a cupcake with no sugar. Having a gold rush to get sugar out of the equation seems like we're missing things from the past. There are many things out there that can do some of the jobs of sugar. It's modern alchemy in every sense. We'll make incremental progress that unlocks other ideas, but turning not gold into gold is very similar to turning not sugar into sugar. There isn't an alternative to sugar.

Much of the future of food in the United States is determined by wealthy white males. I would love a future of food that isn't that. I would love a future of food where instead of eradicating all animal husbandry and growing hamburger slabs in labs, maybe sometimes eat a beet? I would like someone to nail an animal-free, climate-friendly burger, and egg, and chicken so that we can incorporate it into our everyday life so that market forces can refocus on the next thing, which is making fruits and vegetables as good as they can. My dream future of food is that Silicon Valley's hundreds of millions are applied to a reasonable goal: Let's make a really great sweet potato.

The approach right now has been animal muscle, which is biochemically pretty simple. It's like a lawnmower. All of the stuff that a carrot has going on in terms of enzymes and colors, it's like a Ferrari. The interest to date has been how to strip this Ferrari of parts to create a convincing lawnmower. The things a melon can do when it's put through its paces are so far beyond what a steak can do.

Marion Nestle, author of *Let's Ask Marion: What You Need to Know About Food, Nutrition, and Health*, age 84

In twenty years, I hope **food**food** is on our dinner plate. And by that, I mean sustainably grown and raised edible plants and animals, under conditions that promote the health of workers as well as eaters, that are kind to animals, and reduce environmental damage and greenhouse gas emissions. The food challenge for the future is to feed the world's population in ways that promote health and protect the environment, and do so sustainably. Diets that do all this are largely (although not necessarily) plant-based, which for people in industrialized countries means increasing the proportion of food plants and decreasing meat intake.

If I could wave a magic wand I would create a food system that feeds a healthy sustainable diet to everyone on the planet, regardless of income, that pays decent wages to everyone involved in producing, packing, cooking, and serving food, that ensures food safety, and that protects the environment. That's a utopian vision, but I think it's what we need to aim for.

Minh Tsai, owner of Hodo Foods, age 50

Because of how interconnected we are through the web and our travels, our dinner plate in the future will be more global in flavors, with cuisines that include African spices, and Mediterranean flavors like za'atar and ras el hanout. Similarly, Asian flavors like gojuchang and fermented fish pastes will be more prominently used. Taste will remain key to adoption, and nutrition and health will remain the second-most-important decision factor for what we put in our mouths. Does it taste great? Is it good for my health? Is it good for the planet? This is the sequence of questions consumers will ask.

Technical foods' existence and adoption need to serve some purpose. In the foreseeable future, technical foods' primary claim is to save our environment. Why not eat something with comparable nutrition, taste, and economic value to the original foods if it is *better* for the environment? Assuming that tech foods do have a positive environmental impact, the ability to achieve nutritional and economic impact may not be difficult. Yet, we have seen past examples where it's taken time, such as when Daiya came out with meltable cheeses, consumers didn't get excited until Miyoko's Creamery launched with textures that more closely resembled traditional cheeses. Beyond Meat's initial products didn't taste good. Adoption was slow until Beyond fixed its formula and Impossible came along.

I don't think [cultured meat] will become mainstream, even if culinary, nutrition, and price parity are achieved or exceeded. I believe psychologically, we are not ready to have it be mainstream.

I believe and hope that we will eat less of the foods that have detrimental environmental and health impact, whether meat or plants. I am hopeful that the trend of monocrops or CAFOs [concentrated animal feeding operation] will be reduced, simply because they aren't sustainable in the long run. I believe that restaurant chains that are more transparent about their business practices and serve less processed foods will grow faster than the McDonald's and Taco Bells. The younger generation will be more vegetarian and plant-based than the current generation. We will continue to eat less meat, but meat will remain the largest protein source in our diet. Economics dictate choice; people who can' t afford more wholesome foods end up consuming cheaper processed foods, in the United States and globally. This will not change.

I have always hoped for a local food model where you eat produce, fruit, and meats sourced from local growers, or support restaurants that source from local growers. I continue to hope that consumers will care about transparency in the food they consume. There is something wonderful about a sustainable producer that serves a discrete community. That's a win-win.

Sarah Masoni, food innovation expert, Oregon State University, age 56

Our dinner plate will migrate to formal occasions. Most of us will eat foods that are wrapped in edible films that don't need to be prepared and served as we do today. Food will become fuel for the majority of people as the time and ritual associated with meals becomes a special occasion instead of a daily routine. Food as fuel, meaning eating to survive, indicates that indulgence is a thing of the past.

With large populations and limited production, the majority of people will find themselves in a cycle of dreaming of family meals and sitting quietly for a feast as their ancestors did in the past. Survival foods will be nutrient-dense foods that have been preserved with yet-to-be-known technologies, and serviced from in-home vending machines just like we saw on the *Jetsons*. Push the button, you'll hear a whirl, and out pops your food. This system will be managed by a few enormous food companies around the world, and it will be difficult to find a way out of the system for most individuals. For the majority of people, meals will be infrequent, and they will have been created with satiety so that people can go for long periods of time without eating. Our bodies will adapt to this new normal, and the food systems will be fully integrated into our daily lives, not as a hobby or a pleasure center as they are now. Small farms will exist with a counterculture of survivalists who grow and produce their own foods. Underground fortresses, where people are self-contained, already exist today. Agriculture systems that can exist underground with artificial lighting will be sought after by people with money.

Cellular and lab-based meats will become mainstream, and they won't be thought of as strange, they will be without stigma, they will be a necessity. They won't be called lab-based meat, they will be called what they are: steak, or chicken breast.

The basics of our food systems will remain the same. Protein, fat, and carbohydrates will reign as the tenants of calories, but maybe the way that they are arranged will change. Our bodies require nourishment from the three calorie supplies, but the way that they are put together in food may be different. Nutrients that we thought were critical for our survival may become minor, and ones that we thought were less important may become favored. We may not have giant zucchini and undersized apples. Technology may create a perfect growing environment for exact-sized fruits and vegetables for commercial reasons.

Foods from the ocean are emerging as the next platform for foods of the future. Plants that grow rapidly, have nutrient-dense properties, and can be easily made into life-sustaining fuel for humans. I hope for a future of food that can feed the world, because when people are hungry, all sorts of bad things happen. Eating is a survival issue, and hunger is easily the most anti-civilization problem that we have on this planet.

Preeti Mistry, chef/activist, age 44

I'm imagining a more diverse food base. I see a lot of food and flavors from Africa. All cuisines started there. It's the birthplace of civilization, with so many spices and ingredients that have been appropriated. Our palates are craving new things, and the resurgence of BIPOC chefs and cuisines will give us a chance to shine in the foodie world. We'll start opening up our minds and utilizing different vegetables and grains from [diverse] places. And we'll meet that with an ability to farm in a way that's equitable. Folks in other parts of Asia—beyond the standards—will be recognized and paid for their work.

I think cellular meat will become mainstream. I'm not happy about it. I personally think it's gross. It's not even that it's gross; it's sort of the same as plant-based fake meats. Instead of diversifying our diets, most Americans would rather spend money so they can be lazy—putting all this work into something when we could just be more sustainable with our diet choices. I can't believe that people have spent millions of dollars on these products when there are people out there starving. It feels like it's placating Americans because we're lazy babies. Instead of just, "No, you don't get to do that." There are consequences to your decisions. To me it's industrial meat that is expensive and ruining the environment. There are homeless people in Berkeley, and we're trying to make fat in a lab.

What I'd like to believe is that we actually have done something good, and the politicized young people are going to see that we can do things differently, and there is less of Big Ag and Big Org and less "I got my organic salad from Whole Foods that was grown in Chilé, but I feel good about myself because it's organic." In some ways I hope we eat less fast food. I hope we're eating less crap and more local and sustainable. I've seen people young and old who have backyard gardens and are growing stuff. I hope that we continue to see that, and that it will spur more generations to see that there is a different way, and when we do use technology, we use it in a more sustainable way.

On a more theoretical level, I want to see honest good food that's cooked with full intention being given the same value as tweezer food.

As more people get indoctrinated into chef worship and fine dining, we lose sight of what food is. It's sort of like fashion. We look at fine dining as the pinnacle—what we all strive for—and yet it's a thing that is only accessible to a small portion of the world. We look to the folks who feed tiny cakes to the one percent as the innovators and thought leaders. It doesn't make sense. I'm glad we're getting to hit the pause button, and see how it's affected consumers—baking their own bread, growing their own scallions. Maybe they can reconnect [to food] and realize you can't get everything how you want it, when you want it. I hope we recalibrate who's in charge, who we see as leaders, and what we champion.

David Nayfeld, chef/owner of Che Fico, San Francisco, age 37

The pessimist in me thinks nothing changes and it's exactly the same. We will be that much closer to global warming, ruining the Earth, more health issues, more disease and obesity. The optimist in me hopes that we get educated to the fact that animal protein, although a necessity for some folks, is not a necessity every day of the week. At the very least, not three meals a day. The ideal plate in ten to twenty years is 85 percent vegetables, grains, and legumes and 15 percent animal protein. I have a fear that we'll overcorrect until everything is a GMO of something, and overindustrialized, as in the past. I think the right thing to do is take notes from our history and do more plant and crop rotation and eat what the soil provides for us. Even though we are the apex predators, that doesn't mean we should have every whim at the push of a button.

I fear lab-based becomes mainstream, because I believe that everything else we've fucked with has had adverse and unintended consequences. I'm a strong believer that the Earth has put everything we need on it for our survival, and we as human beings are too egotistical to believe that we should be limited. So we try to manufacture every single thing we want.

There's a good chance that some fish will go away—you won't see tuna around for much longer unless we're able to curb our global appetite for them. But they shouldn't be made in the lab. My answer to that is always no. We should all stop eating tuna for a while, and if you want to eat tuna, you know what you should do? Fucking deal with it. You can live without tuna for a while. There are fish we have backed off from. Let's take five years without tuna.

I want people to start eating cover crops like alfalfa, amaranth, and peas. All of those things can make a larger presence in our daily diet. As chefs we can help in popularizing new vegetables, which can in turn benefit farmers. I want to see ancient grains used more and modified wheat used less. I bet if we got off overly processed wheat you would see a shift in people who are experiencing gluten intolerance, and we would see that wheat is not an enemy, but a very sustainable food source.

Nadia Berenstein, PhD, historian of food technology, age 41

I think that what we eat in twenty years will be substantially shaped by climate change. My hope is that responsible food companies will transition to making foods and adopting farming and manufacturing techniques that lessen their impact on the environment. Part of this shift will be driven by consumer demand, part by necessity—but a big push will have to come from government and policy. The huge question for me is how food cost will factor into this transformation. Well-off consumers have already shown themselves willing to pay a premium for "ethical" and sustainable foods. People on a budget or receiving government assistance are rarely able to make such choices. Will we, as a society, commit ourselves to making sustainably grown, nourishing food accessible to all? Will we listen to ordinary eaters, farm laborers, and workers up and down the food chain, and take their needs and expertise seriously?

In the future, we might see a combination of regenerative and sustainable agriculture alongside really high-tech innovations in food production—including synthetic proteins and fats, flavors and flavor modifiers. Rather than be stigmatized, these will be a key part of maintaining a palatable food supply that doesn't just keep us alive, but allows us to maintain a joyous quality of life.

I'm a flavor maximalist. I want to see more varieties of delicious strawberries and more "fake" strawberry flavors—because delicious strawberries may not always be available without a huge environmental or labor cost. I also think there's a future for Big Food here. Big Food is not inevitably in conflict with social and environmental well-being; after all, large companies can marshal technology, innovation, and economies of scale in ways that benefit both eaters and the planet we share.

There's a place for cultured meat—if they're able to solve the numerous production issues. I doubt we'll be seeing lab-grown T-bone steaks or filet mignon on our dinner plates. I think it's more likely that lab-grown animal proteins will serve as components of other protein-rich foods—including foods which have yet to be dreamed of—in combination with other synthetic or natural components. If lab-grown meat can replace some of the meat in a McDonald's hamburger, or some of the chicken in a chicken nugget, the technology will have a more dramatic effect on climate change than if we succeed in making super-expensive slaughter-less sashimi for posh, cruelty-averse diners.

I'm most excited by the idea of ordinary people playing around with home-based bioreactors and other DIY biotech in order to shape their own versions of a synthetic future. I don't think this will ever be the main way people feed themselves, but if biotech and cell culturing turns into a source of play in the kitchen—contributing to culinary social interaction, community building, and pleasure—I would be delighted. People need to feel trust both in the food system and in the government, the entities that ultimately guarantee the safety of what we eat. That's one of the huge problems that still faces us as we move forward. How do we trust the things that are given to us and promised to us as delicious solutions—given that we've been burned many times before? There is no adventure, no common future, without trust.

Tamar Haspel, columnist for the *Washington Post*, age 57

In twenty years our dinner plate is going to look a lot like it does now. I think that foods that are the best candidates for being replaced or rethought are animal-based. I suspect that non-egg-based mayo and non-dairy-based milk are going to be strong between now and then, and I suspect that we will have ever better plant-based replacements for ground meat but not for whole muscle meat. Beyond that I don't think a whole lot will change on that horizon. I think there's a distinct possibility that plant-based meat will be a public health negative. In fact, they have a very similar nutrition profile, people feel virtuous eating them, and once they have that health halo people will overeat it. We've seen it time and time again.

For cell-based to go mainstream, it will really depend on cost. I think it's going to be a long time before it's competitive with plant-based ground beef. Steak is a pipe dream. Until there's a shift away from whole meat muscle demand we're not going to see a decrease in beef herds. The demand for the whole muscle cuts is what drives beef herds now, and I think we're a long way from really affecting animal herds. People care about taste, price, convenience, and healthfulness; everything else is far down the list. For example, people buy organic because they think it's better for them, although its main benefit is environmental. We're going to have people buying plant-based products because they think it's healthier for them, but the more important repercussion is how it plays out in agriculture.

My vision for the backbone of a diet for people and the planet is staple crops. The center of our plate should be whole grains and legumes. They can grow efficiently, they're storable, they can be harvested by machine, they have virtually our entire nutrition complement. If you have grains and legumes as the mainstay of your diet and vegetables and animal as adjuncts, you're doing great.

Soleil Ho, restaurant critic for the *San Francisco Chronicle*, age 33

In terms of climate change, wealth inequality, and equity at large, we haven't answered those difficult questions. In twenty years, I predict that there will be another pandemic and there will be more unrest. The silver lining is that it will shake us out of thinking we can keep kicking the can down the road. It's hard to predict if we reach a tipping point of the way things were done—gratuity, health insurance, a white supremacist stink over everything. As one aspect of life in society, dining will look different, the reason for dining will look and feel different, and the cost will be drastically different.

I don't see cell meat scaling yet. It's also hideous. I think there is still a lot of identity wrapped up in meat eating—in eating a burger from a cow. I wonder if it will become more prolific and if there will be culture wars and [whether] meat has taken on a right-wing meaning. I wonder if lab-based becomes a thing, if it will be ripe for conflict. I think it will be another facet of the culture war.

Part of the question I struggle with is: Who are we talking about? Margarine is gone from Whole Foods, but maybe people still buy it at the dollar store? That's the weird thing. If there are things that are going extinct, then there are people still buying them, and if things become unhealthy, there are people who are marginalized and will still be eating them. We think that food will bring us together, but I hope that there will be more critiques, less of Pollyanna systems thinking when we talk about the future of food. There's a real problem with distribution—some people get the worst food, and some get the best. The future of food is the reckoning and reimagining of hierarchy and wealth distribution, and I hope everyone gets what they need and what they deserve as humans. The future I hope for is one in which we don't have rich or poor people, and where everyone can eat something good and culturally appropriate.

I wonder: If you could just conjure up food, what else could you do with your time? So much of our brainpower has been dedicated to finding food, and it's only recently that we can set it aside so that it's a leisure activity. When I think about the future, it's toward an ideal world. What can lead us to post-scarcity in a way that is equitable—not just for people in a starship or at the top, but for everyone?

Jonathan Deutsch, professor of culinary arts and science at Drexel University, age 44

What we're seeing now are concurrent solutions, which Warren Belasco wrote about in his book *Meals to Come*. There's the tech fix—like cell-based, Impossible, and so on—and the anthropology fix—shifting to plants, more whole grains, and eating lower on the food chain. I don't think either one of those will win. I think we will continue to see these bifurcated approaches. For as long as possible, we are going to try to increase consumption and hedonic satisfaction, which will in all likelihood mean more meat, sugar, and carbs, and more obesity. What makes me think that is looking around the world. All the arrows are going up—we're eating more meat, using more land, we're getting fatter, and we're dying younger. It takes a revolution to change that. Even in this pandemic we're seeing increased food sales. Plant-based sales are up, but meat isn't dead. We're on a trajectory of consumption. Most of the world doesn't have enough to eat or has enough to eat but wants more.

I think cultured meat will be available. I don't think you'll be wandering the supermarket aisle and deciding between conventional and cultured meat, but there is a really interesting locus of opportunity there to grow specific cuts of meat. Where I think this opportunity lies is in the high end. Animals are horribly inconvenient. Why not pay premium prices for cultured-meat cuts? You can go to a premium steakhouse and get a cultured-grown, Wagyu-level, super marbly, eighty-dollar steak relatively quickly, because the margins are there.

We're slow to change. Food habits are very deeply held and ingrained. The most likely thing we'll lose is the diversity of seafood. We're fishing ourselves out of a food supply. Oysters were [once] super abundant and bars wouldn't even charge for them.

I hope for a food system that gets more sustainable, healthier, and more equitable, but I have doubts about the feasibility of that. I also hope for a food system where consumers become advocates for better food. I think the opportunity for food is at the intersection of hedonic pleasure, responsibility—in terms of sustainability and nutrition—and affordability/convenience. This is another great takeaway from Belasco's book, where he has a convenience-responsibility-identity triangle. You can have it fast, you can have it good, or you can have it good-for-you/

good-for-the-Earth. You can probably get two of those things together, but all three are a challenge.

Vaughn Tan, author of *The Uncertainty Mindset: Innovation Insights from the Frontiers of Food*, age 41

What's on our plate in twenty years? I suppose it depends on whose plate the food will be on. The wealthier you are, the better your food choices can be. You know what you want to eat, and you have more buying power. What's likely to be on the plate for most people will be industrialized food: industrially processed from industrially grown ingredients. The usual argument of why industrial is necessary is that food needs to cost less. There are obvious arguments: the public is cash strapped; they aren't exposed to food that is good; and they haven't learned why they should spend more money and time on food.

I think lab meat is one of the stupidest ways to eat meat. There's really only one good reason for eating lab meat, and that's that it didn't come from a sentient being. It seems unlikely to be more energy efficient than meat from a regeneratively farmed animal. It's not more sustainable if you think very broadly about the options we have. You are essentially growing lots of protein in industrial systems. None of this seems to be a good thing to want, or a good outcome. Why not eat meat twice a year? When you really want meat, you can spend the money to buy meat from an animal that was raised properly.

In the future, I don't think we'll be eating incredibly cheap meat. This is a false luxury where everyone who wants to eat meat gets it, but it's full of antibiotics and grown in a feeding facility. The infrastructure needed to raise animals in industrial lots could fall away in the next five years, and that's one reason why so much money is going to lab meat. We'll still be eating industrial meat. We're not getting away from an industrial agricultural system. What we will get away from is the "thing" we're eating.

The future of food I hope for is that people will cook more for themselves with recognizable ingredients. The simple act of cooking most of your food from raw ingredients changes your consumption decisions and the ecosystem of consumption. It creates knock-on effects that are desirable: production and consumption ecosystems that are healthier for the consumer, the producer, the society we live in, and the planet.

I usually never predict the future. But something I've been thinking about is how one important reason why some things are great is that they are not always predictable—they have the ability to surprise and delight us. In order to simulate naturally great food that surprises and delights, the key thing we need to learn to replicate is their unpredictability. No one is thinking about this yet, and I think it's one thing we'll need in the future.

Sophie Egan, author of *How to Be a Conscious Eater*, age 34

There will be a lot more variety than what's on our plates today. In twenty years, the biggest change will be in the number of total species and crops that make up our diets. Rather than purely aspirational diversity, it's necessary to embrace greater biodiversity in agriculture to ensure greater resilience of the food supply in the face of climate change. It's amazing to see the difference in the health of crops grown in healthy soils, and among other species, as well the incredible increase in taste and nutritional value, versus crops grown in soil that has the same thing planted every year.

Cell-based protein will be an option, but it won't be the default. In twenty years, enough types of cell meat will have scaled so that it's affordable and available. I think a decent number of people who eat meat but have reservations about animal welfare will eat it, but I don't think the majority of the population will switch entirely to cell meat. It will become just another option on the menu, so to speak, like the menu of options for protein that's available to us now: pasture-raised, grass-fed, etcetera.

There will be some kinds of food that the planet takes off the table for us. An example of that is palm oil. I'm not saying that we won't have palm oil, but it won't be used in the ubiquity that it's used today because the planet won't support it in terms of deforestation.* We saw the complete collapse of cod, as another example, and we're on track to lose more types of seafood because of overfishing. That will sadly result in some more species from the sea becoming unavailable. The planet will make these hard choices for us.

I hope for a world where the defaults are set up to make it incredibly easy to be a healthy, climate-conscious eater, whereas now you have to go out of your way to be a conscious eater. In the supermarket, you have to be a detective to find foods that fit your personal values—health, animal welfare, supporting local and regional farms, worker welfare, minimizing water and carbon footprints—the whole gamut. My hope for the future is that eating this way is democratized—that the overwhelming ubiquity of choices available and affordable in the marketplace are those that are good for us, others, and the planet. What does that include? Greater biodiversity, lower carbon emissions and water usage, and an agricultural system that produces food that actually nourishes people! And that incentivizes farmers to leave the land better than they found it.

We should also be looking at resurfacing some of the foodways that have proven for generations to be most conducive to physical health, and evidence in nature that they can be produced year after year for generations. Much of this means learning from the wisdom of indigenous foodways. Is there a role for technology? Of course. But a lot of what will be new will really be a rediscovery.

Chef Sean Sherman, founder of The Sioux Chef, age 45

Something we talk about a lot is this complete lack of knowledge of indigenous food and the awareness of these communities' food knowledge. They have a blueprint for living sustainably using plants and animals of different regionally appropriate foods that comes from thousands

* About half of all supermarket products likely contain palm oil – 85 percent of which comes from Malaysia and Indonesia. There has been a 46 percent loss of rain forests in Indonesia from 2003 to 2015.

of years of knowledge. It's called "traditional ecological knowledge," or TEK, which looks at how indigenous people were surviving, how they were eating and processing through their direct connection to the world around them. In twenty years, I hope for a community-based food system, community-based agriculture, a deeper understanding of the wild foods around us, and the ways we can live closer to our environment. We can produce more food if we landscape for it like our indigenous communities. We see this as a positive path for the future—it embraces diversity and brings us closer to our environment.

If you look at popular plant-based products out there, they aren't healthy—they're very high in sodium. As we push forward on future foods, we need to focus more on health and on what makes us healthy. We should be moving away from an American diet where we are over-reliant on animal protein. I think we have a better possibility if we focus on true whole foods that are attainable to people. A lot of the plant-based products can't be made at home. It's an access thing; to make food equity for people, we need to focus on whole foods, community agriculture, ethnobotanical [knowledge], and permaculture design.

I hope we're eating less fast food and convenience food. It's a shift we have to do, especially in America, where our diets are awful. We need to focus on our health, on our land, and on our bodies. I feel like the answer is there, and indigenous knowledge can get us there.

Lisa Feria, CEO and managing director of Stray Dog Capital, age 44

It's going to be more plants, and less conventionally sourced meats. "Meat" made of fungi, plants, cell-based . . . and less of what we eat today. Logistics and distribution sources will be different. There will be a lot more variety in how we get our food—smaller sourced, local farmers, growers outside of the typical system, and greater accessibility for home cooks. There will be a democratization of food and logistics because of COVID-19.

I absolutely think that cell-based will be mainstream in twenty years, from companies that are doing 100 percent cell-based to companies that are doing parts of it to get the plant-based mimicry all the way there on texture, flavor, and taste. Adding just enough cell-based meat to make it delightful, or maybe you need it to be a little more fatty to get the right taste in your mouth. Cultivated can get it there all the way.

Conventionally raised meat and CAFOs are doomed. We will have better prices for healthier foods, versus now, where we have lower prices for unhealthy foods. I see a generational shift. Gen Z and Millennials—the people coming behind us—they already want to eat differently. How can we provide meat that has the cultural value and weight without the negative externalities like disease, deforestation, and pollution? We can use plants that are better for the environment and figure out how to tweak elements to improve them.

In the future, you will be able to customize what you want in your meat, like omega-6. As we are thinking of customized food, we can increase the things that are better for us and decrease the things that aren't—like cholesterol. If we have a cell-based platform for growing parts of animals, why couldn't we grow cells from animals that are no longer around? What if we could eat a dinosaur? Things we never thought were possible! We can deliver things that are beyond our imagination. That's the future I want to hand over. We shouldn't be handing future generations a world on fire because we wanted cheap meat for everyone.

Paul Shapiro, author of *Clean Meat* and CEO of the Better Meat Co., age 41

I think within twenty years, microbial proteins will be a bigger part of our diet than they are today, especially in food ingredients. I believe we can create a much larger amount of protein cheaper this way than by other methods. In twenty years, cultivated meat will be like plant-based is today, which means penetration in supermarkets and fast-food menus, but still a small portion. I think cultivated meat will be available to wide proportions of the mainstream. But I will say that plant-based has been on the market for decades, and still has only one percent of market share. That's a pretty sobering thing to consider, and a helpful reminder to the challenges of scaling up for the future ahead. I think there's a lot that can be done to grow this industry, and blending animal protein with plant protein is a way to help this approach, but we should be cognizant of the facts: Meat consumption has never been higher than it is today.

I believe that many of the cruelest animal products will simply be banned. Numerous states today have already banned the sale of eggs from chickens who are confined to cages, veal from calves locked in crates, sows locked up in gestation crates. There are an increasing number that ban the sale of products that are coming from particularly cruel and inhumane agribusiness practices. I'd like to see a food system that uses vastly fewer resources to produce more food: 1) Less animal suffering. 2) Leaving far greater swaths of land open for wildlife and nature. 3) Preventing world hunger. 4) Reducing emissions and allowing more land to be reforested, so more carbon can be sequestered.

I want to see countertop meat makers in the same way you go to a friend's house and see a bread maker or ice cream maker and it's not remarkable. You'd order tea bags of stem cells and drop them in and make meat. The same way people take weeks to brew beer at home. You could imagine, like, a turducken. What if you could get the cells in one bag and make a turducken, and it would be a culinary experience that no human has ever had? That would be incredible. Another idea that I like to think about is similar to a local bar brewing their own IPA. What if they brew their own meat, and the pig is in the backyard, and you can tip your hat to the pig and eat pulled pork without harming the pig?

Ben Wurgaft, historian and author of *Meat Planet,* age 42

For years, I conducted ethnographic research with people who were quite willing to casually speculate about the future of food. I kept my own speculations to myself, and indeed, I came to think that people' s speculations, and even their formal predictions or forecasts, reflected their aspirations and desires more closely than they did any accurate sense of what's to come. So the only intellectually honest way of answering your question is to be transparent about my own desires. I think we can be fairly certain that climate change will reduce resources—of land, of water—the world has available for farming. This will affect the world unevenly, and different state and corporate actors will try to use their power and influence to feed themselves or maintain profits. We'd be better served by shifting to agricultural strategies that are less intensive, that place less stress on our existing natural resources, and—here's the hard part—we probably need to be a smaller world. I'm on the side of less growth, and especially less population growth. Call it de-growth. In my view, taking the side of continued growth—of markets, of populations—is placing a foolish bet on the continued improvement of agricultural yields or new technological substrates. The pandemic has changed my perspective on our food system—but only by making me more certain than ever that we need fewer production bottlenecks, which is what we've seen at meat processing plants in the United States.

Notes on Sources

This book could not have happened without access to the inner workings of the food-tech world, and sources who talked to me again and again—some going on five years or more. To everyone who has spent time with me, thank you for your patience. And your willingness to join me on the hamster wheel of book publishing.

For years I attended conferences in person. That is where I networked, learned about new companies, and caught up with founders. Once the COVID-19 pandemic hit, I was grounded and my conference life moved online. Virtual conferences mean more people can take part, but they can't replace being there in person. While living in New York City, I started attending the Future Food-Tech conferences and haven't missed one. I attended New Harvest's annual conference in Boston in 2018. I attended the Good Food Institute's annual conference in the Bay Area in 2018 and 2019. Before shelter-in-place orders kept me home, I embarked on my last few research trips, which included MycoTechnology in Denver, Food Tank's annual conference in Manhattan, and AeroFarms in Newark, New Jersey. I moderated a panel on cultured fish at the Cultured Meat Symposium in San Francisco in November 2019. I visited the team at Plenty in November 2019. I flew to San Diego to meet the founders of Plantible and BlueNalu in December 2019. And my last unmasked outing was to Memphis Meats in January 2020. I turned in my manuscript at the end of February, and for two weeks I had the joy of seeing friends and hanging out. Two weeks later the coronavirus pandemic began in earnest, and everything changed overnight. The irony is not lost on me here.

A great many people helped educate me on New Foods. I talked to experts and champions of this New Food movement, including detractors, investors, academics, and founders. They didn't all make it onto these pages, but their time was invaluable. They are, in no special order: Bruce Friedrich at the Good Food Institute, Paul Shapiro at the Better Meat Co., Tara McHugh and Rebecca McGee at the USDA, Brent Loken at WWF, Rachel Cheathem, David Katz and Rachna Desai at Diet ID, Alexandra Sexton at University of Oxford, Babak Kusha at Kilpatrick Townsend & Stockton LLP, Deborah Cohen at the RAND Corporation, Isha Datar at New Harvest, Arvind Gupta at IndieBio, Ryan Bethencourt, Lisa Lefferts at Center for Science in the Public Interest, Amy Rowat at UCLA, Jonathan Deutsch at Drexel University, chef Dan Barber, Kim Severson, Dana Cowin, Kate Krader, Garrett Oliver at Brooklyn Brewery, Christy Spackman at ASU, Emily Contois at University of Tulsa, Asaf Tzachor at University of Cambridge, Stephen Mayfeld at UCSD, Alan Hahn and Josh Hahn at MycoTechnology, Brian Frank at FTW Ventures, Seth Bannon at Fifty Years, Kim Le, Fu-hung Hsieh, Ethan Brown, Pat Brown, Dr. Michael Greger, Michele Simon, Christopher Gardner at Stanford University, nutritionist Ginny Messina, Marion Nestle, Tim Geitslinger, Ryan Pandya and Perumal Gandhi of Perfect Day, Arturo Elizondo and Ranjan Patnaik of Clara Foods, Uma Valeti, David Kay and Eric Schulze at Memphis Meats, Dan Kurzrock, Claire Schlemme and Caroline Cotto of Renewal Mill, Minh Tsai at Hodo Foods, Beth Zotter at Trophic, Elliot Roth at Spira, Sean Raspet at Nonfood, Josh Tetrick and Andrew Noyes at Eat Just, and Tony Martens and Maurits van de Ven at Plantible.

I buried myself in a deluge of books including *Meat Planet* by Ben Wurgaft, *The Third Plate* by Dan Barber, *Meals to Come* by Warren Belasco, *Future Foods* by David Julian McClements, *The Fate of Food* by Amanda Little, *Lentil Underground* by Liz Carlisle, *Clean Mea*t by Paul Shapiro, *Diet for a Small Planet* by Frances Moore Lappé, *The Nature of Crops* by John M. Warren, *Glass Houses: A History of Greenhouses* by May Woods and Arete Swartz Warren, *Food* by Waverley Root, *Mycelium Running: How Mushrooms Can Help Save the World* by Paul Stamets, *The Omnivore's Dilemma* by Michael Pollan, *Billion Dollar Burger* by Chase Purdy, *Slime* by Ruth Kassinger, *Seeds of Science: Why We Got It So*

Wrong On GMOs by Mark Lynas, and *How Not to Die* by Michael Greger. There were more that I've undoubtedly forgotten, including older reference books that I checked out from the Internet Archive (an invaluable resource for anyone doing research—please consider donating to this wonderful nonprofit).

I scrolled through endless studies always with the goal of basing any findings on those that were the most current. I read page after page on the SoyInfo Center's website—an unmatched source for soy history. I read patent applications, food nutrition panels, and ingredient fact sheets.

I did my very best to look at New Foods from all sides, and to balance my reporting with the thoughts of additional experts. Any shortcomings or mistakes are my own.

Acknowledgments

This book began as a seed of an idea that freelance work kept me from watering. Days were hectic: juggling interviews, flying to events and conferences, filing and editing stories, and responding to e-mails. Sometimes I'd fantasize about how nice it might be to work on one single project.

The lure of a book spoke to me because friends and colleagues always seemed to have questions about what they were eating. A book on future foods allowed me to investigate the nooks and crannies I obsess on. Eventually, I dug in and wrote the book proposal. I e-mailed agents, who mostly turned me down. In February 2019, on a trip to São Paulo to attend Fruto, Alex Atala's food conference, my friend Nancy Matsumoto, a journalist I greatly admire, put me in touch with her agent: Max Sinsheimer. I fired off an e-mail to Max from my hotel room. He responded with interest. I am so grateful he saw the value in my project, and me: a first-time author. Max took me on as a client, and then we got down to the real work. With his help, we made my book proposal better until eventually we sold it to Abrams Press. At Abrams, my deep appreciation goes to Garrett McGrath for believing in its potential. My deep appreciation also to a few industry veterans who lent their ears before *Technically Food* became a reality, including Anne McBride, Pam Krauss, Gary Taubes, Dana Cowin, and Laurie Gwen Shapiro.

My mind likes to think the book proposal was the hardest part, but really it was the manuscript. Thank you to the libraries of Marin for being home to my daily efforts, and to the coffee shops that allowed me to sit for hours, including Marin Coffee Roasters in San Anselmo. Thank you to Professor Kent Kirshenbaum for answering my pointed science questions, and arguing with me over cultured meat. For my early readers, the ones who read chapters before they were any good, an especially big thank you, including to my San Francisco writers' group: Zara Stone, Daniela Blei, and

Ellen Airhart. My finished manuscript had the good fortune of falling under Lauren Bourque's review. Along with my inflexible deadlines, Lauren read every page and replied with essay-length e-mails with bulleted lists pointing out where my words were falling short. Then she read it again. For her skills at zooming in and zooming out—on the book and on the words—an enormous thank-you to Sarah Fallon.

I am deeply grateful to so many people who talked to me via phone, Zoom, e-mail, and text during 2020. This included Amy Thompson, Sarah Masoni, Rachel Wharton, Kate Lindquist, and Seth Solomonow. To Kezia Jauron, my favorite animal rights publicist, thank you for being my go-to for all things vegan. To Alan Ratliff for providing me with the most perfect book title, and your endless advice and support throughout, and to Blyth Strachman for showing up and being there throughout this process. To those who aren't listed here but who listened to me rant and inquired after my progress: thank you.

So many read my book in draft, and to you I am fortunate. To Albert Kelly for being a sounding board on many a hike. To Haven Bourque, who read my book on a flight to New York and pointed out where I was missing the bigger picture, or forgetting that this is a global food world made up of communities that walked before me. To Derek Dukes, who told me where I was missing the joke. To my cousin Rafael Zimberoff, who referred to himself in the third person, and introduced me to "Whack," a well-placed reminder that less is more.

Thank you to every start-up that let me sample their foods: AeroFarms, Beyond Meat, Eat Just, Memphis Meats, Perfect Day, Prime Roots, Hooray Foods, Atlast, Spira, Impossible Foods, Meati Foods, MycoTechnology, Plenty, Ripple, ReGrained, Renewal Mill, Pulp Pantry, Clara Foods, Triton Algae Innovations, and New Wave Foods. And to a few of you that weren't quite there yet—Blue Nalu, Aleph Farms and Plantible—maybe one day soon?

Finally, thank you to my family, who put up with never-ending book drone. Thank you for your unflagging support, enthusiasm, and patience.

Index

About the Author

Larissa Zimberoff doesn't like to eat on the go but is known to walk miles to sample something delicious. She is an itinerant fruit forager, and will pluck a white loquat, fuyu persimmon, or green apple off any wandering branch. The story she tells friends will be about scaling the fence, and how great that fruit tasted. College years were spent at UCSD, wandering the eucalyptus groves and beaches of San Diego. Eventually, she landed in the Bay Area. Larissa worked in San Francisco's tech industry for more than a decade, including at Gap Inc, Yahoo!, and Flickr, but found her way out. She moved to New York in 2011, lived on the Lower East Side, and earned an MFA in creative nonfiction from the New School in 2013. She's written on the promise of fake eggs, the potential of peas, and how artificial intelligence saved a winery. Her work has appeared in the *New York Times,* the *Wall Street Journal, Wired, Time, Bloomberg Businessweek,* and many more. Her passion for technology is matched by an equally precious hope for foods that comes from the natural world. She lives in Northern California, where she's close to mountain hikes and seaside cycling. This is her first book.